Environmental Change in South Asia

Anup Saikia · Pankaj Thapa
Editors

Environmental Change in South Asia

Essays in Honor of Mohammed Taher

Springer

Editors
Anup Saikia
Department of Geography
Gauhati University
Guwahati, India

Pankaj Thapa
Kanglung, Bhutan

ISBN 978-3-030-47659-5 ISBN 978-3-030-47660-1 (eBook)
https://doi.org/10.1007/978-3-030-47660-1

This Springer imprint is published by the registered company Springer Nature Switzerland AG
The registered company address is: Gewerbestrasse 11, 6330 Cham, Switzerland

Preface

Prof. Mohammed Taher (1931–2015)

Mohammed Taher (1931–2015) was born in Khutakatiya, a village near Lakhimpur, a town in Assam, India. He went on to study at Cotton College in Guwhati, take his Masters in Geography from Calcutta University in 1958 and receive a Ph.D. from the University of Auckland on a Commonwealth scholarship in 1965. At Auckland, he wrote a dissertation entitled 'Asians in New Zealand: a geographical review and interpretation' under the supervision of Dr. F. H. A. G. Zwart and Prof. K. B. Cumberland. His work focussed on Chinese, Indian and Lebanese immigrants to New Zealand and their descendants. The thesis, of 492 pages, was completed in a short span of time of 2 years, an indication of his diligence and academic and intellectual abilities.

Professor Taher was a teacher par excellence and a meticulous researcher and served Gauhati University in various capacities, beyond his duties to the Department of Geography. At a recent visit to Gauhati University in September 2021, Dr. Rajkumar Ranjan Singh, Union Minister of Education and External Affairs, Government of India, who received his Ph.D. under Prof. Taher, pointed out that when 'Taher Sir' taught students something, it was almost never forgotten.

Professor Taher was variously Rector, Gauhati University, Acting Director of the Omeo Kumar Das Institute of Social Change and Development and Chairman, Board of Secondary Education, Assam. When entrusted with any responsibility, his diligence and mettle was consistently unwavering.

Professor Taher was a noted and very well respected scholar in academic circles in northeast India, well beyond the discipline of geography, equally at home with anthropology and sociology, perhaps attributable to the interdisciplinary nature of his doctoral research and to association with close friends such as Dr. A. C. Bhagawati and Dr. M. S. Prabhakar in those fields. Professor Taher was also a prolific writer and authored some 30 odd books, both in Assamese and English.

As an individual, he was humble, unassuming and always ready to help and advice a colleague or student alike. These attributes were evident in his dealings with people outside the university's limits as well: little wonder then that Prof. Taher was extremely popular and widely respected throughout the state of Assam and northeast India. His popularity and locus standing in Assamese society was evident, when on his passage on February 25, 2015, the Assam Tribune, a widely circulated newspaper, carried the news as a bold headline on its front page, bemoaning the loss of this noted citizen.

This collection of papers is a small token of respect to the scholarly and genteel Prof. Mohammed Taher.

Guwahati, India — Anup Saikia
Kanglung, Bhutan — Pankaj Thapa

Contents

Chapter 1
Developing a Sustainable Cradle for Humanity to face Global Environmental Challenges—The Case of South Asia

Sneha Rani Deb and Priyanka Sengupta

Abstract South Asia has been a canvas of urbanization, industrialization, modernization and westernization without a properly chalked out plan, thus resulting in situations with unintended consequences. The environment in this ecologically sensitive zone has been cornered, the repercussions of which fall onto humanity in the form of disasters and a gradual degradation of social, political, economic and biological milieu. This chapter presents the threats of environmental and climate change faced by South Asia and strategies contrived by global thinkers and change-makers and their applicability to the South Asian context. Poverty, increasing fragility of states from a biodiversity perspective, overshoot issues, simplification of food webs, decrease of biomass per trophic level and ecocide are all reflections of degradation of ecology via climate change. In the 1970s, scientists, engineers, geographers, ecologists, economists and biologists pondered on ways to turn the tide on climate anomalies. In due course of time, they adopted a multilateral perspective to try to control the looming clouds of global warming, GHG emissions, erratic weather phenomena and imbalances in natural systems. What we are now witnessing is the outcome of pushing the ecosystem towards the edge, as humanity itself stands at the precipice. Corporate greenwashing has hindered the pace of efforts directed towards ecological sustainability through efforts like circular economy, carbon capture utilization and storage.

Keywords Unintended consequences · Fragile state · Overshoot · Ecocide · Corporate greenwashing · Circular economy

1.1 Introduction

If the entire history of the Earth is compressed into a 24 h time frame, then humanity came into the picture only in the last 77s. Within this sliver of time, human behavior has brought about unfathomable changes on the face of Earth. Such changes include

S. R. Deb (✉) · P. Sengupta
Department of Geography, Gauhati University, Guwahati, India
e-mail: snehadeb435@gmail.com

A. Saikia and P. Thapa (eds.), *Environmental Change in South Asia*,
https://doi.org/10.1007/978-3-030-47660-1_1

loss of forest cover, de-glaciation, desertification, habitat loss, frequent extreme events, sea water intrusion far inland, species extirpation and extinction, dwindling ecosystem services, etc., which degrade the quality of life, and also hinder the fulfillment of basic human rights like right to clean water and air. It is mostly the marginalized communities who bear the brunt of human-induced climate change. A narrow view of economic development has thus resulted in collapse of environmental thresholds and it has also posed a serious challenge to human welfare. In order to ensure long-term development that is able to self-sustain through positive feedback loops, it became necessary to look at development from a sustainable viewpoint. The first leap towards sustainability was achieved in the form of Brundtland Report (Our Common Future), by the World Commission on Environment and Development (WCED) released in 1987. The report defined sustainable development as "development that meets the needs of the present without compromising the ability of future generations to meet their own needs." (*Report of the World Commission on Environment and Development: Our Common Future*, 1987; UNESCO, n.d.)

It is now an established fact that the extent of changes that have already come about cannot be completely reversed. In order to make the Earth's environment congenial for the continued survival and proliferation of all life forms, sustainable measures are needed. Humans as the wisest species on Earth are at the forefront of these attempts at renewal and regeneration of the Earth's biosphere. Such attempts are still at a natal stage and hence, we have considered them to be sustainable cradles for humanity.

South Asia, is rich in biodiversity but at the same time has the ecological potential to cause adverse effects to the atmosphere, hydrosphere, lithosphere and biosphere, locally and globally. These occur as a result of disruptions in the self-regulatory mechanism of the natural world; disruptions ushered in anthropogenic activities. In the cities of the global south, adverse effects to the environment materialize not merely from deforestation and land use change (LUC). Instead, negative ecological effects are ascribable to the synergies of 'unintended consequences' (Norton 2016). The lure of materialistic eudaimonia and consumerism has had adverse fallouts on the environment.

> In July 2010, torrential rains that hit Pakistan were unprecedented in the northern highlands. The Indus and its tributaries inundated its levees to wreak havoc, affecting 20 million people. Even before the rains appeared in the scene, deglaciation had already ballooned the flow of Indus, thus making the floods of Pakistan a climate catastrophe (Brown 2011). In mid-October 2020, forest fires in Uttarakhand, India were equally unprecedented and 12 times more than those experienced in previous years. This can be attributed to the climatic disorientation which trailblazes neoteric heat waves, sweeping over the Himalayan states (Sajwan and Singh 2021).

Humans have always been the toreadors in taking on the environment, but the route to moral existence lies in understanding and respecting the subtle balances and synergistic interactions between humans, other species and the fragile environment. The environmental degradation-poverty linkage is well known and dealt with at length elsewhere (Olanipekun et al. 2019; West et al. 2014). The shackles of poverty weigh down people to the realms where they face the brunt of environmental cataclysms. What do we call the present scenario—a plenipotentiary environment bearing the burden of the Meghalayan age or, the environment sending humanity encrypted messages of a more calamitous future?

Consider the case of Maldives and Bangladesh:

> 2014, Maldives: this archipelago battles increased rainfall, cyclones and storm surges and severe erosion due to rising sea waters where hundreds of islands report erosion, many of which are severe in nature. The political leaders of this pristine landscape have sought international funding to build sea-walls as a last resort. With 530,000 people exposed to storm and cyclones, 80% of Maldives lies just 1m above MSL (Pal and Ghoshal 2020) and though it barely contributes to global warming it faces the consequences drastically (Schafer 2019).

> Bangladesh: All thanks to the climate change issue that considerable Bangladeshis are subjected to the ill-effects of intruding saline waters, flash floods, storm surges and drought, resulting in the bottlenecking of Dhaka. The last decade saw the internal displacement of 700,000 people (McDonnell 2019; Internal Displacement Monitoring Centre). Records also speak of innumerable flocks of 'paperless' people walking across the borders of India at the dead of the night (Tripathi 2016).

In the light of these cases and similar ones occurring across South Asia, appropriate climate action is of utmost necessity. In this paper, we have noted the pressing environmental issues and their scales of operation in South Asian context. An attempt has also been made to explore grass-root level efforts directed towards sustainable cradle designs and their feasibility. Concurrently, the forces acting as a leash in sustainability efforts have also been studied and articulated.

Environmental Dynamite

The economic indicators that mankind has been resorting to while weighing development since one and a half century present a panoramic view of men and economy. However, such indicators fail to address the issues of ecology. As a result, many countries that seem to be on economical pedestals might have pressing ecological issues. Such ecological issues remain masked as environmental dynamites. It is only a matter of time that the ecological thresholds give off and an economy that seems flourishing today, succumbs to the ecological injuries. A few of such environmental dynamites that are quite prevalent in South Asia are discussed below.

(i) Low income and rising food prices—As a rule of thumb, developing economies are characterized by low-income families struggling to eke out a living in the face of rising food prices. This ghoulish romance between the two, pushes forward social evils in the form of food shortages followed by food riots, and the upwelling of dirty food politics (Brown 2011) and unfortunately the "highest burden of hunger in absolute terms is to be found in Southern Asia" (FAO 2015).

(ii) Hunger sprawl—The rise in grain prices and imbalances between economic development and food scarcity, has hit the densely populated regions of South

Asia particularly hard. With governments offering food subsidies, the absolute numbers of hunger stricken has substantially decreased, but hidden hunger remains unabated. Hidden hunger is defined as the insufficiency of micronutrients, or, infinitesimal uptake of vitamins and minerals. This leads to malnutrition even though food intake by an individual may be sufficient.

(iii) Fragile states—Akin to a crumb of radioactive element that emits life threatening radiations, a fragile state or a failing state spews corruption and terrorism to states near and far. It can be defined as a spatial expanse that was once a self-governing, sovereign state but the government has, presently under economic impoverishments broken down to the point where it is unable to provide basic services to its people (Ryan and Lumley 2000). The failure of the government does not owe its roots only to socio-political disruption, but also to environmental failure. The web of state fragility, environment and climate change, violence, economy and development, societal cohesion and repulsion are not easily understood; the linkages are intricate and alarming. It has been pointed out by the International Institute for Sustainable Development (IISD) that climate change would hasten the sagging trajectory of fragile states (International Institute for Sustainable Development 2011). Unpredictability and unpreparedness for environmentally disruptive forces, increased vulnerability of communities, unplanned and unsustainable development measures are drivers poised to convert states into fragile states (Table 1.1).

Fragility specifies a sui generis rank given to each country, based on various indicators covering economic, political and social conditions and cohesive factors prevailing within a state. The rank being aggregate is hypersensitive to its constituent factors, leading to pushing up or pulling down of a state's rank due to slight changes. The lower the rank is, the greater is the state's vulnerability to climate change. It is to be noted that the ranking is a political risk management and not in itself an ultimatum, thus, the ranks are an early warning for the international geo-political community (The Fund for Peace 2005).

Table 1.1 Ranks of countries (2021) based on fragility

Name of the country	Rank
Afghanistan	9
Bangladesh	39
Bhutan	96
India	66
The Maldives	99
Nepal	51
Pakistan	29
Sri Lanka	55

Source www.fragilestatesindex.org/

(iv) Water-based food bubble—With concretization of landscapes proliferating across South Asia, and indeed across much of the globe, water levels have fallen drastically. Today, much of the agriculture is carried out through unsustainable and injudicious means, following which the bubble of water used inflates (Brown 2011). This results in over-pumping of water from Earth's treasury. Most of the irrigation done in the alluvial plains of India and China comes from over-pumped waters. The process of cultivation goes on well until irrigation is constant. However, once water shortages appear, the productivity decreases conspicuously, thus bursting the food bubble. In this way, the depletion of aquifers and the receding ground water levels lead to food shortages. Erratic rainfall patterns result in drought or famine- like conditions that are problematic to the health of fund strapped developing nations. Sharing of water resources across international borders is already a burning issue amidst countries like India and China; such problems will spark new debates in the upcoming decades (Nair 2020).

(v) Biotic homogenization—Biotic homogenization is an ecologically deleterious process where different biotic communities become similar and analogous over time; a few invasive or strong species winning over weaker ones (Rosenblad and Sax 2016). The downward flow of diversity is due to disturbances and disorganization in the gene pool caused by the turmoil of natural activities (viz. asteroid impact, drifting of landmass, volcanic eruption, earthquakes, tsunamis, etc.) or human activities (from deforestation to habitat losses and LUC). Throughout Earth's history, biotic homogenization took place in several chapters. But the rate at which anthropogenic activities cause species extinction is beyond comparison (Barnosky et al. 2011; Olden et al. 2004). Prehistorically, wherever *Homo sapiens* set foot, they extirpated local fauna, for example, Australian mega fauna (45,000 years ago), Neanderthals (30,000 years ago), American mega fauna (16,000 years ago), *Homo floresiensis* (13,000 years ago) (Harari 2015). The scenario is quite troubling now; humans have exacerbated climate change, global trade and reckless hunts (Richardson and Pyšek 2007) and the fallout is not encouraging. Biotic homogenization simplifies ecosystems, making species vulnerable to extirpations. The most prominent example of extirpation includes *Trachypithecus Geei* from the Peacock Island, Assam, India (Parashar 2020). Global warming creates a vicious cycle of size reduction, early senescence and distribution of species. Body metabolism, in a warm environment happens at a rapid rate causing early onset of sexual maturity and a proportionately small body size. With predators' body sizes unaffected, the number of hunts by a predator needs to increase exponentially. In the time frame of a few decades, specialist species are erased from the landscape, leading to a wide-spread distribution of generalist species (Lurgi 2020).

(vi) Rampant consumerism and waste maximization—Mahatma Gandhi's famous quote "The world has enough for everyone's need, but not enough for everyone's greed" is relevant in an age of rampant consumerism. Consumerism has embedded itself in the psyche of people as divisions between wants and needs are blurred with the passage of time. With the rise of consumerism, demand and increased output have accelerated raw material, energy, and water usage as waste generation has mounted. Developed countries have exported consumerism to the developing countries though free trade and have found in highly populated South Asia a ready market. Synthetics such as polyester, nylon, rayon, spandex or elastane, etc. that are non-biodegradable and e-waste have mounted. With modern lifestyle come modern food choices and packaged items. A staggering amount of waste generated in this process and has conjured up waste management issues. Producing less waste is an obvious solution, although with growing numbers and affluence in South Asia as elsewhere, this remains elusive.

(vii) Plastic Pollution—Plastic which breaks down into micro-plastic, reaches the bodies of all organisms through food web linkages. All mobile creatures other than humans do not realize when or how plastics enter their food bubble and consume the same unknowingly. The food web then distributes the micro-plastic to other trophic levels. Micro-plastics have been found even in human placenta (Ragusa et al. 2021). The toxic properties of plastic tamper with our natural bio-chemical mechanisms and have far-reaching health impacts. Single-use plastics are a serious threat and ecocide or the intentional or unintentional destruction of natural ecosystems through human actions, has entered into the equation. However, 'destruction', in this case points to a situation that has aggravated beyond repair. Humans are now choking ecosystems as human wants continue to burgeon. Consumer requirements, when multiplied by global population, create a huge impact beyond the comprehension of individual consumers and ripple effects of mismanagement have made life precarious for natural systems.

The unscrupulous use of scientific innovations coupled with population explosion in the South Asian countries, has increased the sphere of the above-mentioned issues. Although stated as issues of South Asian countries, it does not rule out the possibility of other nations being vulnerable to similar issues. Ecosystems do not follow political territoriality, neither in bestowing resources nor in afflicting cataclysms. And, hence, there exists cause–effect relationship between the issues and the global environmental offsets.

Measuring the Scale of Environmental Offsets

Quantification of environmental offsets bear great significance for countries to adequately gauge the enormity of environmental issues and to plan effective adaptation and response strategies better suited to their local ecology. There are numerous means at our disposal for measuring the scale of environmental offsets that have been around for quite some time. Some of the most relevant measures are discussed below.

(i) Earth Overshoot Day—"Earth Overshoot Day marks the date when humanity's demand for ecological resources and services in a given year exceeds what Earth can regenerate in that year. We maintain this deficit by liquidating stocks of ecological resources and accumulating waste, primarily carbon dioxide in the atmosphere. Earth Overshoot Day is hosted and calculated by the Global Footprint Network, an international research organization that provides decision-makers with a menu of tools to help the human economy operate within Earth's ecological limits. To determine the date of Earth Overshoot Day for each year, Global Footprint Network calculates the number of days of that year that Earth's biocapacity suffices to provide for humanity's Ecological Footprint. The remainder of the year corresponds to global overshoot. Earth Overshoot Day is computed by dividing the planet's biocapacity (the amount of ecological resources Earth is able to generate that year), by humanity's Ecological Footprint (humanity's demand for that year), and multiplying by 365, the number of days in a year" (Earth Overshoot Day and Global Footprint Network).

This can be expressed as follows:

(Earth's Biocapacity/Humanity's Ecological Footprint) $\times$ 365 $=$ Earth Overshoot Day.

The further the overshoot day in a calendar year, the better chance we have at succeeding in bringing back ecological balance in the environment. The earlier the overshoot day, the more severe is the skew in ecological balance that occurs.

Earth Overshoot Day has been on a declining trend since 1970 with minor fluctuations (Table 1.2 and Fig. 1.1). For the year 2020, a landmark year for drastic reduction in human activity and resurrection of the natural conditions, courtesy of the Covid-19 pandemic, Earth Overshoot Day fell on August 22, slightly later than the date in 2019, i.e., July 29. The Global Footprint Network calculated and consolidated a reduction of 9.3% in the Global Ecological Footprint in 2020 compared to 2019. The two major sectors that contributed to this reduction are dialing down of wood harvest and reduced CO_2 emissions from the combustion of fossil fuels (Earth Overshoot Day and Global Footprint Network). This delay in overshoot day points to the possibility of preserving our Earth's natural state if conscious lifestyles are adopted. By 2020, the world's population was using up resources that could be supplied by 1.6 Earths. The #MoveTheDate movement has gained traction as more global citizens are now paying attention to their personal ecological footprint and putting in efforts in whichever ways possible using mediums at their disposal.

Overshoot days may also be calculated for countries. A country's overshoot day represents the date of the Earth Overshoot Day, if the world population lived like the people of that country. Figure 1.2 depicts the country overshoot days for selected countries for the year 2020. The worst performing countries in this regard are Qatar and Luxembourg, whereas the best performing countries are Indonesia and Ecuador. The world community needs to follow the ideal countries so as to cut down on their environmental costs and strive towards achieving some semblance of ecological balance. Most developed countries have their overshoot

Table 1.2 The dates of past Earth Overshoot Days (based on the National Footprint and Biocapacity Accounts 2019)

Year	Earth Overshoot Day	Year	Earth Overshoot Day	Year	Earth Overshoot Day
1970	December 29	1987	October 23	2004	September 1
1971	December 20	1988	October 15	2005	August 25
1972	December 10	1989	October 11	2006	August 19
1973	November 26	1990	October 11	2007	August 14
1974	November 27	1991	October 10	2008	August 14
1975	November 30	1992	October 12	2009	August 18
1976	November 16	1993	October 12	2010	August 7
1977	November 11	1994	October 10	2011	August 4
1978	November 7	1995	October 4	2012	August 4
1979	October 29	1996	October 2	2013	August 3
1980	November 4	1997	September 29	2014	August 4
1981	November 11	1998	September 29	2015	August 5
1982	November 15	1999	September 29	2016	August 5
1983	November 14	2000	September 23	2017	August 1
1984	November 6	2001	September 22	2018	July 29
1985	November 4	2002	September 19	2019	July 29
1986	October 30	2003	September 9	2020	August 22

Source www.overshootday.org/newsroom/past-earth-overshoot-days/

days in the first half of the year. This signifies the scale of ecological footprint they generate as compared to the developing countries.

(ii) Anomalies in ecological budget–Anomalies in ecological budget may occur in negative or positive directions. An ecological debt occurs when the environmental resources are used beyond the Earth's natural renewing capability. A negative environmental or ecological budget leads to an ecological deficit or debt. Although humans as a species are responsible for this deficit, the human population in its entirety cannot be held accountable for the scale of ecological debt that we owe to our Earth as resource exploitation and consumption patterns are heavily skewed in different areas of the Earth. The blame game is quite strong in this case. The developing or third world countries claim that rich or first world countries "have systematically expropriated their natural resources for profit, either without paying at all, as in the case of bio-piracy of genetic plant, animal and human resources, or by paying too little" (Simms 2001). The brunt of ecological imbalance and climate change is faced by the developing countries which are not remotely responsible for the present state of affairs. This unfair distribution of resource use and climate change impacts has caused the developing world to ask the developed world for compensation. Developed countries, on the other hand, wish to coerce developing countries

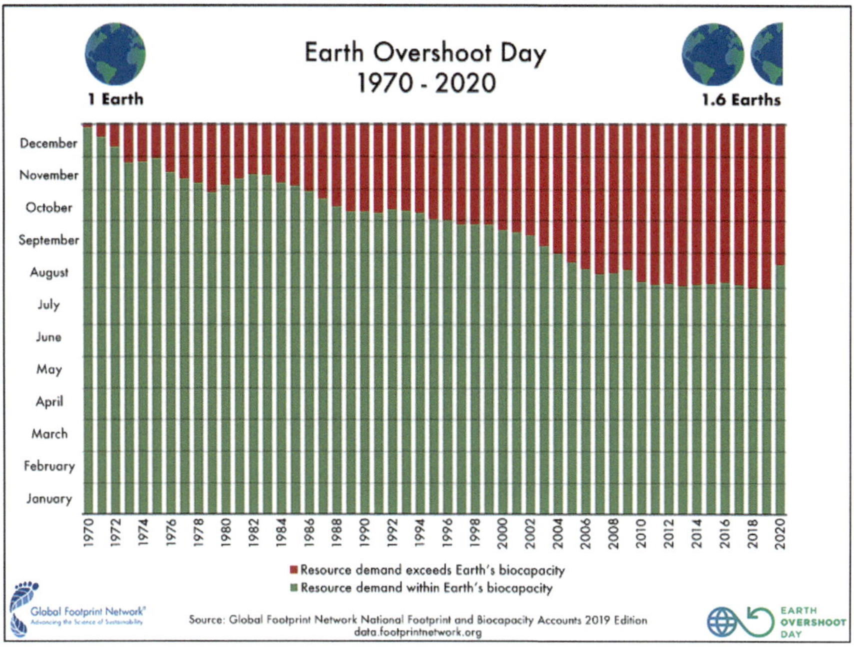

Fig. 1.1 Earth Overshoot Day (1970–2020)

> into slowing down economic development with reasoning that the latter are responsible for more wasteful and unsustainable development compared to the former. However, they miss out an important point. Developed countries are now able to invest in cleaner energy or greener initiatives only by means of their wealth that they accumulated following the same path now trodden upon by developing countries. "Who owes whom" (Simms 2001), albeit significant for fair sharing of compensation efforts, is not the prominent question in the third decade of the twenty-first century. It is necessary to look beyond country-wise accountability and delve deeper into personal accountability so that each global citizen makes conscious decisions every day for a better Earth.

Whether a country is an ecological debtor or creditor depends on their ecological footprint and biocapacity. When the ecological footprint of a country exceeds its biocapacity, it is deemed as an ecological debtor. In a reverse case, the country becomes an ecological creditor. "An ecological deficit occurs when the Footprint of a population exceeds the biocapacity of the area available to that population. Conversely, an ecological reserve exists when the biocapacity of a region exceeds its population's Footprint. If there is a regional or national ecological deficit, it means that the region is importing biocapacity through trade or liquidating regional ecological assets, or emitting wastes into the global commons such as the atmosphere. In contrast to the national scale, the global ecological deficit cannot be compensated for through trade, and is therefore equal to overshoot by definition" (Global Footprint Network).

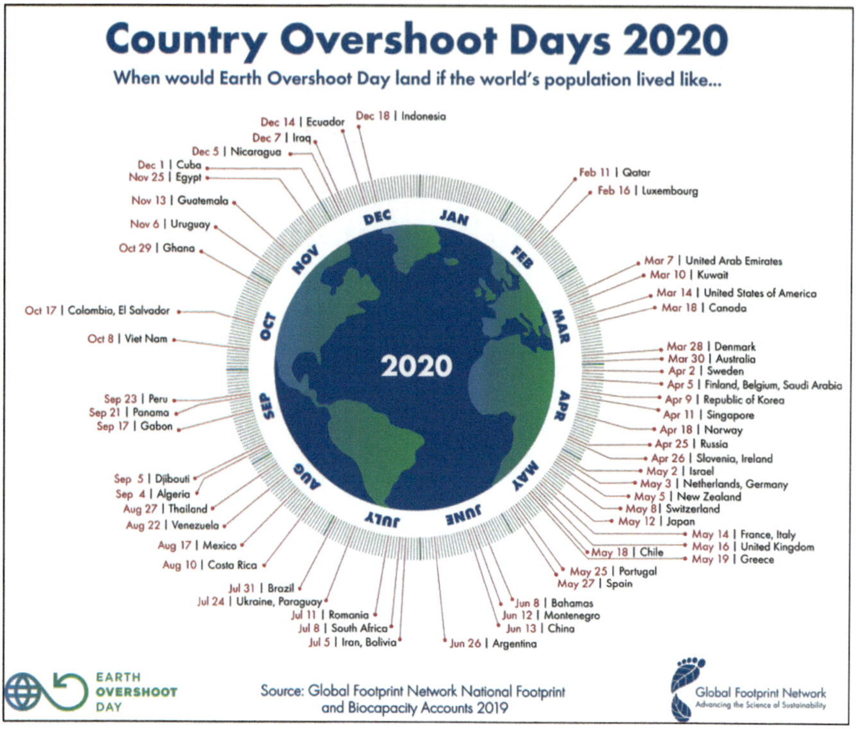

Fig. 1.2 Country Overshoot Days (2020)

South Asia, as a whole, is an ecological debtor in terms of global hectares (i.e. biologically productive hectare) as its overall ecological footprint exceeds its overall biocapacity (Fig. 1.3). However, in terms of number of Earths required to sustain a world population which lived like the people in South Asia, the picture is quite promising. Less than 1 Earth is required if the global population lived like the South Asian population (Fig. 1.4). This implies ecological surplus and bounties of positive feedbacks.

Room for Optimism?

The year 2250 BCE set the rhythm of deformation for the world's first urban center in Uruk, Mesopotamia. The city boasted intelligence, craftsmanship, agriculture and controlled a population of around 10 million. Unfortunately, human skill could not interpret the warning of the ecosystem and the severe drought that set in. The faulty irrigation system and the insolation together depleted the soil moisture, turning the soil into a heap of mineral salts.

Environment has always been the keynote of sustenance on Earth and the ebb and flow of civilizations are dictated by environmental quality. After the Orbis Spike (Biello 2015; Lewis and Maslin 2015) of 1610 AD, colonialism, global trade and

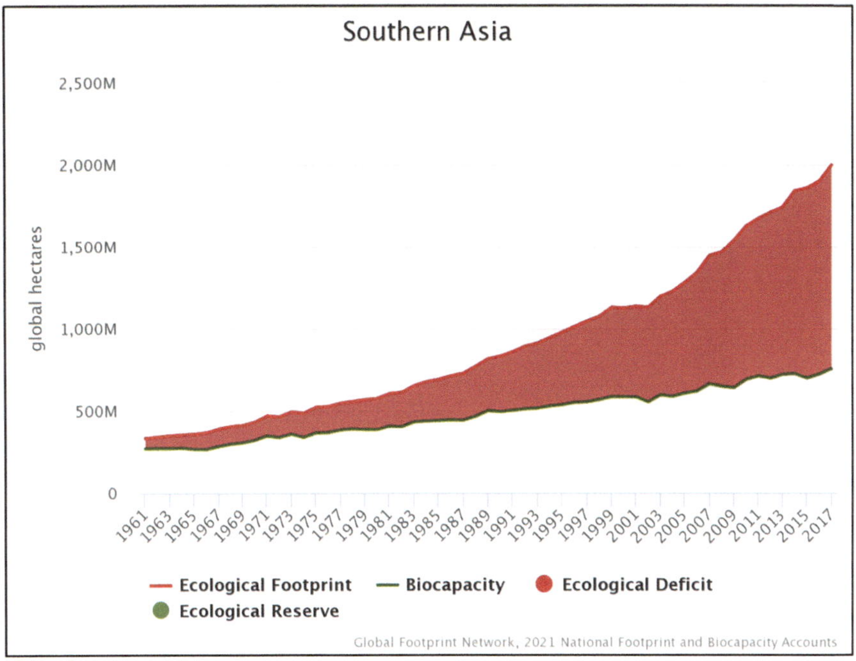

Fig. 1.3 Ecological Footprint versus Biocapacity (gha) for South Asia

the fossil fuel coal initiated humanity's first serious impact on the Earth. The good news is that humanity has started to realign the thoughts from being 'an esurient' to 'a shepherd' of nature. There still exist ways and means to combat climate change and stem further deterioration of environmental quality, a few of which are discussed below:

(i) Circular economy—"A circular economy is a systemic approach to economic development designed to benefit businesses, society and the environment. In contrast to the 'take-make-waste' linear model, a circular economy is regenerative by design and aims to gradually decouple growth from the consumption of finite resources" (Ellen MacArthur Foundation 2017). The circular economy model can be applied to all areas—food, fashion, travel, lifestyle, etc. When applied to food, the circular economy basically includes buying food without packaging like fresh fruits and vegetables not wrapped in plastic, composting food scraps and using it as manure for kitchen gardening, growing food, if possible, not wasting food by either cooking food in required proportions or by feeding left-overs to strays or beggars.

Circular economy model applied to fashion has given birth to circular fashion. Circular fashion involves not only the recycling of clothing and other fashion items but also allied spheres such as use of natural and minimally processed materials which, at the end of their life, are able to biodegrade easily, i.e., without

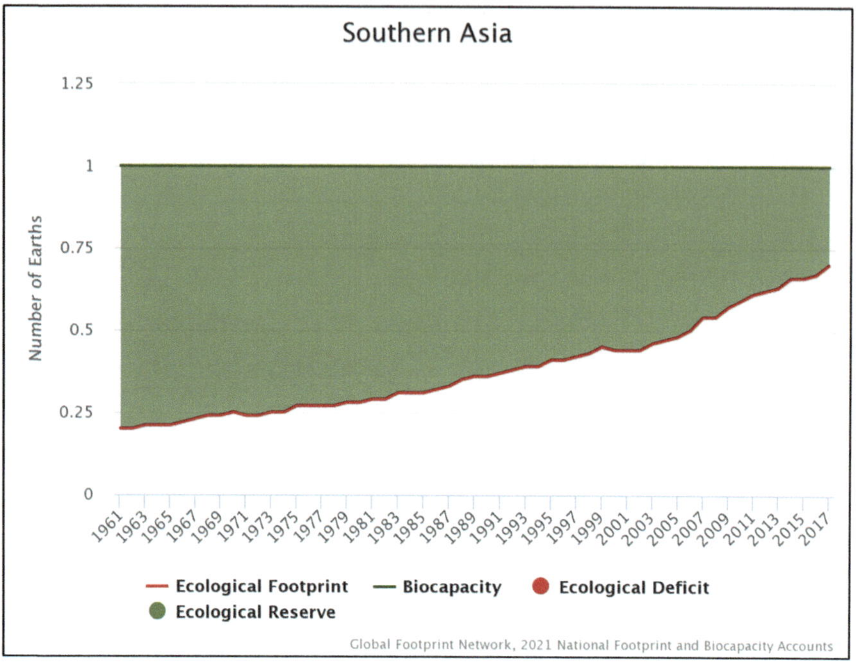

Fig. 1.4 Ecological footprint (number of earths) for Southern Asia

much stress on the Earth's biodegrading mechanism; use of natural dyes which make products safe for health of the consumer as well for the environment; eliminating fast fashion and going about slow fashion (Carus and Dammer 2018). When it comes to natural materials for clothing, the age-old pure cotton is the obvious winner. Clothing brands now have a '100% organic cotton' collection, which rules both the comfort and sustainability games. Next in line is linen, which is made from flax plant, coarser than cotton but equally sustainable. Wool is also a natural fiber and therefore, easy on the environment. Silk is yet another natural fabric made from silkworm cocoons. All other fabrics are not eco-friendly, no matter what we are made to believe. Since South-East Asia has a suitable environment for textile growth—including availability of materials mentioned above, a booming market for natural fibers as opposed to western markets, skilled artisans, etc.—sustainable and bio-degradable clothing is easily available and affordable.

Pinatex is a natural sustainable material made from waste pineapple leaf fibre. The raw material used is a by-product of agricultural activities performed by pineapple cultivators (Ananas Anam, n.d.). AppleSkin™ is another natural material made from the waste of fruit juice industry. The skin and the core of the apple fruit are made of cellulose fibers, the waste materials (raw materials in this case) that go into production of AppleSkin (Luxtra London 2022). These products are made by upcycling agricultural and food waste and are, therefore, accurate examples of the cradle to cradle design concept. These plant-based materials are worthy leather alternatives. Many brands worldwide are now embracing these materials to climb up the sustainability ladder. Given South Asia's agricultural potential, such products can be produced efficiently to minimize the environmental impact of conventional leather and uplift subsistence farmers.

(ii) Incentives—Incentives or rewards are great ways to support and propel change makers. One such reward for sustainable fashion in India is the Circular Design Challenge (CDC). It is "a platform for young fashion/accessory designers and entrepreneurs to showcase and win a prize for their innovative ideas and collections made by using materials from diverse waste sources including plastic. Launched in collaboration with Fashion for Earth by R|Elan™, UN Environment and Lakmè Fashion Week, it is one of India's first and largest sustainable fashion challenges in the fashion, textiles and apparel industry". The winner of CDC 2020 was 'Malai' (Fashion for Earth by R|Elan™). 'Malai' is an Indian brand that makes goods out of *malai*—"a newly developed biocomposite material made from entirely organic and sustainable bacterial cellulose, grown on agricultural waste sourced from the coconut industry in Southern India". This material mimics traditional leather and is reasonably strong, flexible, breathable, water resistant, recyclable, biodegradable, vegan and sustainable. Until now, the material has been used to make bags, backpacks, wallets, pouches, footwear and upholstery (Malai). Work is being carried on to stretch its applicability.

(iii) Responsible tourism—Circular economy in travel and tourism is a bit harder to actually practice, given the mold of the system we are in. Similar to slow fashion, there is the concept of slow travel. It includes cutting down on air travel as the energy used and CO_2 released in the process is huge, choosing an eco-friendly mode of transport such as public transport over private, electric vehicle over vehicle running on fossil fuels, train or bike over cars, walking or cycling for shorter distances, etc. There is now a buzzword in tourism known as 'staycation'. As opposed to vacation, staycation implies staying at home during holidays and engaging in leisure activities at home or going on day trips to nearby places. Visiting local places rather than far-off overseas places cuts down on carbon footprint.

(iv) Cradle to cradle designs—Cradle to cradle (C2C) designs are biomimetically devised by Michael Braungartand it conflates products, manufacturing processes, industry, and businesses to decrease toxins released into natural systems. These designs seek to increase the scope of ecological footprint rather than the wastes left behind. C2C designs comprise of two interlinked cycles that work together, wherein residues left from the technosphere are recycled and treated to become food (nutrition) for the biosphere (EPEA 1990; Hattum 2007). An assessment body under C2C designs, known as SGS has its operating branches in many of South Asian countries, viz., Afghanistan, Bangladesh,

India, Pakistan and Sri Lanka, where products are assessed to optimize and suit environmental standards (SGS 2016).

(v) Carbon Capture, Utilization and Storage (CCUS)—Carbon Capture, Utilization and Storage (CCUS) is a crude yet efficient method to combat the accumulation and amalgamation of carbon in the natural systems. Enhanced Oil Recovery is a technique, which countries in South Asia ought to follow in the coming decades. The Indian Oil Corporation (IOC) has collaborated with the US-based Dastur International to build its largest project in Gujarat (Gupta Uma 2021). Zero carbon emissions, carbon neutrality are some peripheral scopes which can help in reducing the accumulation of carbon in the atmosphere. Maldives, with its low carbon development strategy, plans to curb GHG emissions between 1.3 MtCO2e and 1.5 MtCO2e by following a strict abatement curve, though it is yet to be seen in practice (Villy 2014). Indigenous innovations like that of a plant in Thoothukudi, India, captures CO_2 from its furnace to make baking soda, need to be supported (Trivedi 2020).

Combating Climate Change: Lagging Pace

Although numerous ideas, concepts, technologies are at our disposal, most of them are in infancy. The stages of development, application, practicability and implementation of such combat methods need to pass through several trials and inspections. Cultural barriers, market dynamics, consumer behavior and preference, resource availability and transferability, entrepreneurship, etc., determine the path and pace of a sustainable product from ideation to execution.

(i) Human behavior—The gross domestic product of the South Asian region is highly skewed due to the vast disparities in the distribution of wealth and services. At one extreme lies the poverty stricken with limited access to basic amenities, while at the other lies the super-rich with access to excess of services. Also, the variation in cultural groups is vast, rendering possible varied and differentiated uses of resources. With the spike of visibility of goods, products and services in South Asia, the urge to buy commodities at competitive prices also increases. This causes a cycle where companies manufacture substandard products, often at the cost of environmental quality. Even environment conscious individuals and groups are caught in a loop they cannot seem to break free from. They are caught in a constant state of eco-guilt. Once an individual comes to terms with the scale of environmental impact a decision can have, it can never be unlearnt. Thus, whenever a situation demands an inappropriate action that may harm the environment, eco-conscious people naturally develop an anxiety. Eco-guilt or eco-anxiety is the guilt people experience when they could have adopted a sustainable option but opted not to for some reason.

(ii) The Valley of Death—International organizations, communities, institutions, panels and volunteers call out for innovation, ideas, skill, methods, products and commodities that might help in "net zero carbon emissions". Unfortunately, whatever solutions come up for handling climate change issues come along with an expensive price tag in terms of at the minimum, at least a onetime investment. For instance, solar cells, wind turbine accessories, direct air capture technologies, non-toxic dyes, etc. Due to their price tags, investors and several countries do not participate in their manufacture. As a result, the price cannot be reduced and made available to common people and thus, the company suffers the loss of manufacturing the initial loss. This trough wherein most innovations get stuck is called the Valley of Death. However, government subsidies and loans are known to pull out significant inventories out of such scenarios (Zwilling 2013).

(iii) Corporate Greenwashing—Businesses masquerading as eco-friendly to boost sales are called greenwashing. Examples include printing 'eco-friendly', or a picture of a plant or leaves on the packet or opting for a green color. Companies brainwash consumers into believing that they are really 'green', i.e., they are causing least harm to the environment in their production process or are giving back to the environment through afforestation or other means. Greenwashing has come into vogue in present times due to increased demand for eco-friendly products among the conscious consumers of today. People are really putting efforts to look for products that use natural ingredients which are able to bio-degrade when disposed-off after use, or products made from recycled materials, or products which when purchased, contribute towards plantation activities. Some products mention taglines such as 'made from 100% biodegradable materials', 'made from recycled plastic', 'for each product you buy, we plant a tree'. However, not all of these claims are actually authentic and consumers are often tricked into buying products from businesses that actually don't practice what they preach. They are analogous to a wolf in sheep's clothing. Such charlatans are often busted but some continue to lurk and render conscious consumer choices futile. Greenpeace launched its Stop Greenwash campaign in 2009 to bring attention to the fact and make consumers aware of such frauds (Scientific American 2013).

Humans have become the sculptors, designers as well as modifiers of the Earth and its environment, all the while vouching for short-term gains and overlooking the long-term repercussions of unsustainable routes to progress. However, the need of the hour is not a single, sustainable practice that promises to work with cent percent efficiency, but a pluralistic set of environmental ideologies that incorporate holistic ideals such as eco-modernism, urban political ecology, anarcho-primitivism and green politics. It is through an intricate network of such ideologies that a stable sustainable system be built for human sustenance coupled with optimum environmental quality. Indeed, South Asia with its teeming millions and hence increased pressures on its already burdened ecosystems must prioritize environmental conservation before it is too late.

References

Ananas Anam (nd) About us—Piñatex. https://www.ananas-anam.com/about-us/. Accessed 27 June 2022

Barnosky AD, Matzke N, Tomiya S, Wogan GOU, Swartz B, Quental TB, Marshall C, McGuire JL, Lindsey EL, Maguire KC, Mersey B, Ferrer EA (2011) Has the earth's sixth mass extinction already arrived? Nature 471(7336):51–57. https://doi.org/10.1038/nature09678
Biello D (2015) Mass deaths in Americas start new CO_2 epoch. Sci Am. https://www.scientificamerican.com/article/mass-deaths-in-americas-start-new-co2-epoch/
Brown L (2011) World on the edge. W.W. Norton & Company, pp 4–5, 59–60. https://doi.org/10.4324/9781849775205
Carus M, Dammer L (2018) The Circular bioeconomy—concepts, opportunities, and limitations. Ind Biotechnol 14(2):83–91. https://doi.org/10.1089/IND.2018.29121.MCA
EPEA (1990) Cradle to cradle—rethinking products—EPEA. EPEA. https://epea.com/en/about-us/cradle-to-cradle
Ellen MacArthur Foundation (2017) The circular economy in detail. https://archive.ellenmacarthurfoundation.org/explore/the-circular-economy-in-detail
FAO (2015) The state of food insecurity in the world. http://www.fao.org/3/i4646e/i4646e.pdf
Harari YN (2015) Sapiens: a brief history of humankind. Penguin Random House, p. Timeline of History
Internal Displacement Monitoring Centre (nd) *Bangladesh|IDMC*. https://www.internal-displacement.org/countries/bangladesh. Accessed 31 May 2021
International Institute for Sustainable Development (2011) Climate change and fragile states: resilient development and the struggle for security. International Institute for Sustainable Development
Lewis SL, Maslin MA (2015) Defining the anthropocene. Nature 519(7542):171–180. https://doi.org/10.1038/NATURE14258
Lurgi M (2020) Live fast, die small: how global heating is simplifying the world's ecosystems. The Conversation. https://theconversation.com/live-fast-die-small-how-global-heating-is-simplifying-the-worlds-ecosystems-144976
Luxtra London (2022) AppleSkinTM|Vegan leather made from apples—LUXTRA. https://luxtralondon.com/pages/apple-skin
McDonnell T (2019) Climate change creates a new migration crisis for Bangladesh. National Geographic, Environment. https://www.nationalgeographic.com/environment/article/climate-change-drives-migration-crisis-in-bangladesh-from-dhaka-sundabans
Nair SK (2020) Climate change and water: ripple marks. Down To Earth. https://www.downtoearth.org.in/blog/water/climate-change-and-water-ripple-marks-69829
Norton R (2016) Unintended consequences. In: Econlib, vol 48, no 4, pp 441–444. https://doi.org/10.1057/s41279-016-0005-5
Olanipekun IO, Olasehinde-Williams GO, Alao RO (2019) Agriculture and environmental degradation in Africa: the role of income. Sci Total Environ 692:60–67. https://doi.org/10.1016/J.SCITOTENV.2019.07.129
Olden JD, Poff NL, Douglas MR, Douglas ME, Fausch KD (2004) Ecological and evolutionary consequences of biotic homogenization. Trends Ecol Evol 19(1):18–24. https://doi.org/10.1016/j.tree.2003.09.010
Pal A, Ghoshal D (2020) We can't wait': Maldives desperate for funds as islands risk going under. Reuters, p 2. https://www.reuters.com/article/us-climate-change-maldives-idUSKBN1ZG0XS
Parashar U (2020) 36 years after they were first introduced, Golden Langurs disappear from Umananda Temple|Hindustan Times. Hinduatan Times. https://www.hindustantimes.com/more-lifestyle/36-years-after-they-were-first-introduced-golden-langurs-disappear-from-umananda-temple/story-eLdI4jh0mSh8BQz8uazPVN.html
Ragusa A, Svelato A, Santacroce C, Catalano P, Notarstefano V, Carnevali O, Papa F, Rongioletti MCA, Baiocco F, Draghi S, D'Amore E, Rinaldo D, Matta M, Giorgini E (2021) Plasticenta: first evidence of microplastics in human placenta. Environ Int 146:106274. https://doi.org/10.1016/J.ENVINT.2020.106274
Report of the World Commission on Environment and Development: Our Common Future (1987). https://sustainabledevelopment.un.org/content/documents/5987our-common-future.pdf

Richardson DM, Pyšek P (2007) Elton, C.S. 1958: The ecology of invasions by animals and plants. London: Methuen. Progr Phys Geogr 31(6):659–666. https://doi.org/10.1177/0309133307087089

Rosenblad KC, Sax DF (2016) A new framework for investigating biotic homogenization and exploring future trajectories: oceanic island plant and bird assemblages as a case study. Ecography 39(July):1–10. https://doi.org/10.1111/ecog.02652

Ryan JM, Lumley J (2000) And finally. failed states and failing states. Trauma 2(3):231–236. https://doi.org/10.1177/146040860000200307

SGS (2016) Cradle to Cradle|SGS India. Cradle to Cradle. https://www.sgsgroup.in/en-gb/consumer-goods-retail/hardgoods/furniture/furniture-certification/cradle-to-cradle

Sajwan R, Singh M (2021) Climate change is real: six months on, Uttarakhand forests still ablaze. Down To Earth. https://www.downtoearth.org.in/news/climate-change/climate-change-is-real-six-months-on-uttarakhand-forests-still-ablaze-76318

Schafer H (2019) Bracing for climate change is a matter of survival for the Maldives. End Poverty in South Asia. https://blogs.worldbank.org/endpovertyinsouthasia/bracing-climate-change-matter-survival-maldives

Scientific American (2013, July 29). https://www.scientificamerican.com/article/greenwashing/. Accessed 28 May 2021

Simms A (2001) Ecological debt—balancing the environmental budget and compensating developing countries. World Summit on Sustainable Development. https://wedocs.unep.org/handle/20.500.11822/8232

The Fund for Peace (2005) Methodology|Fragile States Index. Fragile States Index. https://fragilestatesindex.org/methodology/

Tripathi S (2016) Illegal immigration from Bangladesh to India: toward a comprehensive solution—Carnegie India—Carnegie Endowment for International Peace. Carnegia, India. https://carnegieindia.org/2016/06/29/illegal-immigration-from-bangladesh-to-india-toward-comprehensive-solution-pub-63931

Trivedi V (2020) Carbon capture technology not on track to reduce CO_2 emissions. Down To Earth. https://www.downtoearth.org.in/news/climate-change/carbon-capture-technology-not-on-track-to-reduce-co2-emissions-74718

UNESCO (nd) Sustainable development. https://en.unesco.org/themes/education-sustainable-development/what-is-esd/sd. Accessed 24 Apr 2022

Uma G (2021) India's largest industrial carbon capture and utilization project. Pv Mag Int. https://www.pv-magazine.com/2021/02/18/indias-largest-industrial-carbon-capture-and-utilization-project/

Villy J (2014) Maldives Low Carbon Development Strategy. UNEP Risø Centre

West PC, Gerber JS, Engstrom PM, Mueller ND, Brauman KA, Carlson KM, Cassidy ES, Johnston M, MacDonald GK, Ray DK, Siebert S (2014) Leverage points for improving global food security and the environment. Science 345(6194):325–328. https://doi.org/10.1126/SCIENCE.1246067/SUPPL_FILE/WEST.SM.PDF

Zwilling M (2013) Survive the valley of death. Forbes. https://www.forbes.com/sites/martinzwilling/2013/02/18/10-ways-for-startups-to-survive-the-valley-of-death/?sh=36afd51769ef

van Hattum R (2007) Waste is food—VPRO documentary. YouTube.https://www.youtube.com/watch?v=4pwCFH1LkCw

Chapter 2
Assessment of Climate Change Impacts on Floods and Low Flows of the Brahmaputra River

A. K. M. Saiful Islam, Khaled Mohammed, and G. M. Tarekul Islam

Abstract The Ganges-Brahmaputra-Meghna (GBM) river system plays a key role in the survival and development of more than 670 million people in South Asia. The extreme flows of the GBM rivers also dictate the occurrences of floods and hydrological droughts in Bangladesh, which lies at the delta of this river system. This study was undertaken to assess the impacts of high-end climate change on the extreme flows as well as the mean monthly flows of these rivers at their downstream locations inside Bangladesh. SWAT Hydrological modeling tools were used to simulate future flows using climate projections collected from the CORDEX initiative. The mean monthly flows are likely to increase in most months of the future in the GBM rivers, and the increases are likely to be largest in the Ganges River compared to the other two rivers in terms of percentage changes. Flood flows and low flows are projected to increase in all three rivers. The frequency of occurrence of flood flows is likely to increase and that of low flows are likely to decrease, especially near the end of this century. The projections presented in this article can be useful in adaptation planning as well as in supporting discussions on mitigation policies.

Keywords Climate change · Extreme flows · Ganges-Brahmaputra-Meghna (GBM) basins · SWAT model

2.1 Introduction

In conducting climate change impact assessments of river basins, the following modeling chain is usually adopted. First, climate models are used to project future meteorological variables for the desired years, typically for a duration of more than 30 years. Thereafter, these projected data are preprocessed, namely, by downscaling

---

A. K. M. Saiful Islam (✉) · K. Mohammed · G. M. Tarekul Islam
Institute of Water and Flood Management (IWFM), Bangladesh University of Engineering and Technology (BUET), Dhaka 1000, Bangladesh
e-mail: akmsaifulislam@iwfm.buet.ac.bd

G. M. Tarekul Islam
e-mail: tarek@iwfm.buet.ac.bd

A. Saikia and P. Thapa (eds.), *Environmental Change in South Asia*,
https://doi.org/10.1007/978-3-030-47660-1_2

and bias correction. Finally, these preprocessed data are used as forcing to different hydrological models that are set up over a selected river basin to simulate hydrological variables for the future. In this chapter, all these elements of the modeling chain are briefly discussed, followed by an example of application of this modeling chain to assess the changes in future flows of the Brahmaputra River due to climate change. Climate models, also known as General Circulation Models (GCMs), are basically numerical weather prediction models that are run for a long time over a global domain at climatic scales of 30 years or more. Another difference of climate models from weather models is that while the atmospheric forcing of the weather models must be values as observed in reality, in the case of climate models, they can be different from the real-world scenario to investigate the response of the climate system to those different values. Since future forcings cannot possibly be known, a range of estimated values based on various possible scenarios are used to project future climates. A particular component of interest in these possible scenarios is the concentration of greenhouse gases in the atmosphere, which is the prime stimulant behind global warming. In the latest Fifth Assessment Report (AR5) of the Intergovernmental Panel on Climate Change (IPCC), four Representative Concentration Pathways (RCPs) were defined based on four sets of socio-economic assumptions, namely, RCP2.6, RCP4.5, RCP6.0, and RCP8.5 (van Vuuren et al. 2011). RCP2.6 represents the lowest amount of global warming while RCP8.5 represents the highest.

The most common climate models at present are the Atmosphere–Ocean General Circulation Models (AOGCMs) which couple the atmosphere with the ocean, land, and sea ice. An advancement over the AOGCMs is Earth System Models (ESMs) which further includes various biogeochemical cycles such as the carbon cycle, nitrogen cycle, or the sulfur cycle. There are many GCMs available at present that have been developed in countries from all over the world. About 39 models participated in the Coupled Model Intercomparison Project Phase 5 (CMIP5) (Taylor et al. 2012). Before running these models to project future climates, their parameters are calibrated in such a way so that they can reliably simulate the historically observed climate. The better a climate model can simulate the climate of the past, the more reliable it is assumed to be for simulating the climate of the future. A major problem in using the outputs of GCMs for hydrological predictions is their coarse resolution. A typical GCM can have a horizontal resolution of 2°, which is about 200 km near the equator. That means the outputs for a particular meteorological variable are provided as the spatially averaged values over an area of about 200 × 200 km. Obviously mesoscale processes like precipitation vary widely within a short span of location and so reasonable projections of hydrological variables cannot be expected using precipitation data that is spatially averaged over such a large area. This is where the concept of downscaling comes into play, which can convert these coarse resolution GCM outputs into finer scale information. Downscaling can be done either statistically or dynamically. Statistical downscaling simply relates the GCM outputs of the historical period with locally observed data using a statistical function and then uses this function to convert GCM outputs of the future at a local scale. Stationarity of the relationship between the GCM output and the local observation in the future period is therefore an inherent assumption of this process. On the other hand, dynamical downscaling is performed using Regional Climate Models (RCMs), which are very similar to GCMs. The difference is that the horizontal resolution of RCMs is very high (around 25–50 km or even less) and instead of running the models over the whole world, the RCMs are runover a limited area using lateral boundary conditions derived from the GCMs. Some of the most commonly used RCMs include the U.S. Regional Climate Model Version 3 (RegCM3), UK Met Office Hadley Centre's Regional Climate Model Version 3 (HadRM3), German Regional Climate Model (REMO), and the European Centre-Hamburg (ECHAM) model. Even though RCMs reduce the horizontal resolution of the climate simulations, simulated variables such as temperature and precipitation often show significant systematic biases. Using these values to simulate hydrological variables are likely to propagate the errors into hydrological simulations. That is why RCM outputs are almost always bias corrected before being used as forcings to hydrological models. Several bias correction methods are available which vary in complexity. Similar to statistical downscaling, bias correction methods assume that the relationship between the RCM outputs of the historical period and the observed values will remain stationary in the future periods. Some of the commonly used bias correction methods are linear scaling, power transformation, variance scaling, delta change correction, and quantile mapping (Teutschbein and Seibert 2012). Hydrological models are used to represent the hydrologic cycle and simulate its various components. The typical inputs required by these models are meteorological data such as precipitation, temperature and relative humidity, topographical information, soil information, land use/land cover information, and values of several parameters describing the hydrological processes of the study area. Hydrological models can generally be classified into three types: empirical models, conceptual models, and physically based models. Empirical models are data-driven models which are, as their name suggests, based on empirical relationships between various components of the hydrological cycle. Conceptual models use semi-empirical equations and have lumped parameters for describing hydrological processes. Physically based models use complex mathematical equations to calculate the different hydrological variables and use values of spatially distributed parameters. Examples of hydrological models include the Soil and Water Assessment Tool (SWAT), Variable Infiltration Capacity (VIC), Water–A Global Assessment and Prognosis (WaterGAP), and Joint UK Land Environment Simulator (JULES) (Kauffeldt et al. 2015).

2.2 The Study Area

The Brahmaputra is a transboundary river that has an annual average discharge of approximately 20,000 m^3/s (Jian et al. 2009), making it the fourth largest river in the world in terms of average discharge. Its drainage area of 520,000 km^2 encompasses China, India, Bhutan, and Bangladesh (Immerzeel 2008). It originates in southern Tibet of China and travels about 2900 km through China, India, and Bangladesh before ending in the Bay of Bengal (Gain et al. 2011). The climate of the northern

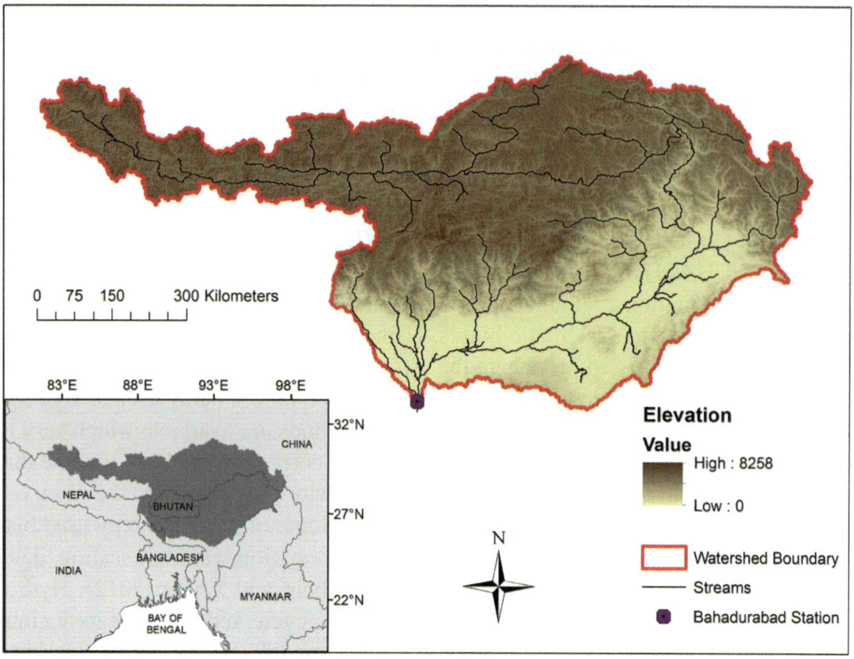

Fig. 2.1 Location and details of the Brahmaputra River Basin

part of the basin that is located over Tibetan Plateau with elevations above 3500 m is cold and dry. The remaining parts of the basin are mostly located on the low-lying floodplains with elevations below 100 m. This southern part has a warm and humid tropical monsoon climate. The mean annual precipitation in the basin is around 2300 mm and about 60–70% of this falls in the monsoon season (June to September). The Brahmaputra River Basin is shown in Fig. 2.1.

2.2.1 *Database and Methodology*

Daily precipitation and temperature data were collected from 11 different climate projections. The reason for using an ensemble of projections in climate change assessments is that the climate models available at present often disagree with one another in the projected values of different variables at different locations. Therefore, by using an ensemble of projections instead of a single projection, the uncertainties that are inherent in the different GCMs and RCMs can be partly accounted for. The 11 selected projections were generated using 10 GCMs of CMIP5 and later dynamically downscaled using 3 RCMs by the Coordinated Regional Climate Downscaling Experiment (CORDEX) (Giorgi and Gutowski 2015). A list of the projections are

Table 2.1 CORDEX—South Asia climate projections used in the study

Institute	GCM	RCM
CSIRO	ACCESS1.0	CCAM-1391M
CSIRO	CCSM4.0	CCAM-1391M
SMHI	CNRM-CERFACS-CNRM-CM5	RCA4
CSIRO	CNRM-CM5	CCAM-1391M
SMHI	ICHEC-EC-EARTH	RCA4
CSIRO	MPI-ESM-LR	CCAM-1391M
MPI-CSC	MPI-M-MPI-ESM-LR	REMO2009
SMHI	MPI-M-MPI-ESM-LR	RCA4
SMHI	NOAA-GFDL-GFDL-ESM2M	RCA4
SMHI	IPSL-CM5A-MR	RCA4
SMHI	MIROC-MIROC5	RCA4

given in Table 2.1. All the projections have a horizontal resolution of 0.5° and were bias corrected with a Multi-segment Statistical Bias Correction (MSBC) method as described in Grillakis et al. (2013). The MSBC method used here is of the family of quantile mapping correction methods. The reference dataset used for the bias correction was the WFDEI dataset (WATCH Forcing Data methodology applied to ERA-Interim data) (Weedon et al. 2014).

The topographic information of the area was collected in the form of a Digital Elevation Model (DEM), namely, the hydrologically conditioned version of the Shuttle Radar Topography Mission (SRTM) DEM of 90 m resolution from the HydroSHEDS database of the United States Geological Survey. A global land use/land cover map of 300 m resolution called GlobCover prepared by the European Space Agency for the year 2009 was collected and the Digital Soil Map of The World prepared by the Food and Agriculture Organization of the United Nations was collected as soil information. Finally, observed discharges of the Brahmaputra River at Bahadurabad gauging station were collected from the Bangladesh Water Development Board (BWDB) for the years 1980–2009. The location of the station is shown in Fig. 2.1. The Soil and Water Assessment Tool (SWAT) was used as the hydrological model. SWAT is a physically based, semi-distributed, watershed-scale, computationally efficient, continuous-time hydrological model that operates on a daily time step. It divides a basin into sub-basins by overlaying a land use/land cover map, a soil map, and a DEM. The sub-basins are further divided into lumped units called hydrologic response units (HRU) which are the percentage of a sub-basin area that has a unique combination of soil, land use/land cover, and slope properties. Using moisture and energy inputs provided by the user, the model then predicts the hydrology at each HRU using a water balance equation which consists of daily precipitation, runoff, evapotranspiration, percolation, and return flow components. The generated flow of all the HRUs in a sub-basin is then summed together and routed through the channels, ponds, and reservoirs to the basin outlet. Detailed descriptions of the model can be found in Arnold et al. (1998).

Using the collected DEM of the study area, the automatic watershed delineation command of SWAT defined a stream network and delineated the outline of the complete basin given the location of the basin outlet (Fig. 2.1). The SWAT model was run at a daily time step for all purposes, i.e., for calibration, validation, and simulation of future discharges. The first 20 years (1980–1999) of the observed discharge data were used for calibration and the remaining 10 years (2000–2009) were used for validation. Calibration was done for only the SWAT parameters that were found to be the most sensitive to Brahmaputra River's discharges by a separate tool called SWAT-CUP (Calibration and Uncertainty Program). Before simulating the future discharges of a particular climate projection, the SWAT model was calibrated and validated using the baseline period of that same climate projection. The Sequential Uncertainty Fitting II (SUFI-2) algorithm of SWAT-CUP was used for calibration.

2.3 Results

To analyze future river discharges, four time slices were considered. These are the baseline period (1980–2009), the 2020s (2010–2039), the 2050s (2040–2079), and the 2080s (2080–2099). The mean monthly discharges of the Brahmaputra River in these time slices are shown in Fig. 2.2 as boxplots. Each box includes data from all the 11 climate projections. Large uncertainties can be seen in the projections. Based on the change in median values of these boxes, the months from March to July will see an increase in mean monthly discharge for all future time slices. The months from September to December will see a decrease in mean monthly discharge for all future time slices. The largest increase for all three future time slices is in March, with values of 41, 86, and 147% during the 2020s, 2050s, and 2080s, respectively. The largest decrease for all three future time slices is predicted in December, with values of 13, 28, and 39% during the 2020s, 2050s, and 2080s, respectively.

Parametric frequency analyses were performed on the annual maxima and minima of the simulated discharges. The Generalized Extreme Value distribution and the Weibull distribution were used to fit the maxima and minima datasets, respectively. The return period curves as estimated by the parametric frequency analysis performed on the annual maxima are shown in Fig. 2.3. Four shaded regions, one per time slice, show the range of return periods estimated by the 11 climate projections. The uncertainty range in the annual maximum discharge is seen to increase with the return period. The solid lines show the ensemble means. Based on the ensemble means, the annual maximum discharges at different return periods are predicted to increase during the 2020s, 2050s, and 2080s compared to the baseline period and the increase is slated to be highest during the 2080s. For instance, the annual maximum discharge with a 100-year return period will increase by 47% during the 2080s compared to the baseline period.

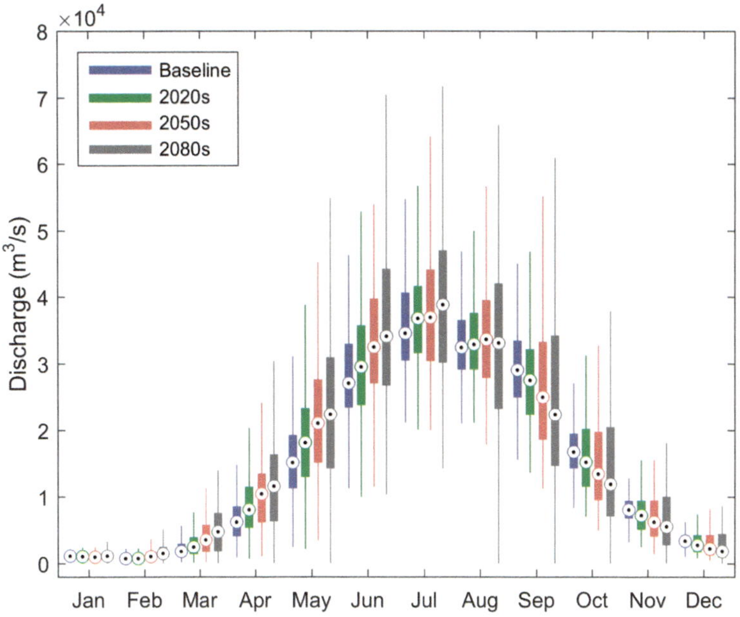

Fig. 2.2 Mean monthly discharges of the Brahmaputra River at different time slices

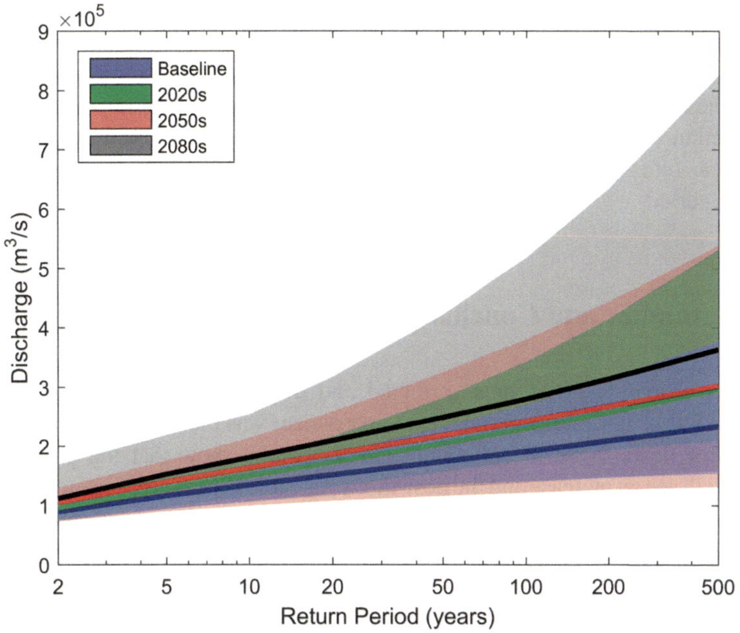

Fig. 2.3 Return period curves of annual maximum discharges at different time slices

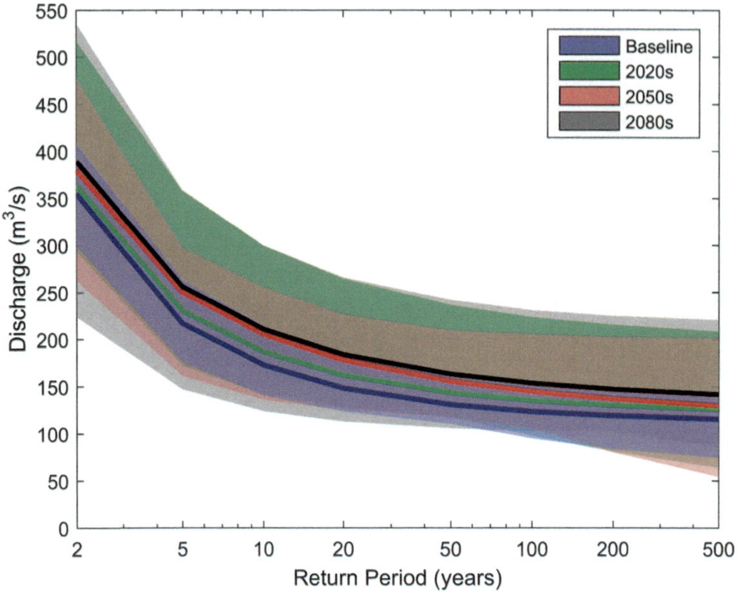

Fig. 2.4 Return period curves of annual minimum discharges at different time slices

Similarly, the return period curves as estimated by the parametric frequency analysis performed on the annual minima are shown in Fig. 2.4. Based on the ensemble means, the annual minimum discharges at different return periods will increase during the 2020, 2050, and 2080s compared to the baseline period and the increase is maximum during the 2080s. For instance, the annual minimum discharge with a 100-year return period will increase by 24% during the 2080s compared to the baseline period.

2.4 Summary and Conclusions

Using an ensemble of 11 bias corrected and downscaled climate projections to force the SWAT hydrological model, an assessment was made of the possible future changes of flows in the Brahmaputra River. Results show that the pre-monsoon months will see an increase and the post-monsoon months a decrease in mean monthly discharges for all future time slices. The month of March has been predicted to register the largest increase and December the largest decrease in mean monthly discharge compared to the other months. By the end of the century, floods are likely to become more frequent in the future and their magnitude is slated to become more severe. Low flows are projected to become less frequent in the future and their magnitude is likely to become less severe.

References

Arnold JG, Srinivasan R, Muttiah RS, Williams JR (1998) Large area hydrologic modeling and assessment Part I: Model development. J Am Water Resour Assoc 34(1):73–89

Gain AK, Immerzeel WW, Weiland FCS, Bierkens MFP (2011) Impact of climate change on the stream flow of the lower Brahmaputra: trends in high and low flows based on discharge-weighted ensemble modelling. Hydrol Earth Syst Sci 15:1537–1545

Giorgi F, Gutowski WJ Jr (2015) Regional dynamical downscaling and the CORDEX initiative. Ann Rev Environ Resour 40:467–490

Grillakis MG, Koutroulis AG, Tsanis IK (2013) Multisegment statistical bias correction of daily GCM precipitation output. J Geophys Res Atmos 118(8):3150–3162

Immerzeel WW (2008) Historical trends and future predictions of climate variability in the Brahmaputra basin. Int J Climatol 28:243–254

Jian J, Webster PJ, Hoyos CD (2009) Large-scale controls on Ganges and Brahmaputra river discharge on intraseasonal and seasonal time-scales. Quar J Roy Meteorol Soc 135:353–370

Kauffeldt A, Wetterhall F, Pappenberger F, Salamon P, Thielen J (2015) Technical review of large-scale hydrological models for implementation in operational flood forecasting schemes on continental level. Environ Model Soft 75:68–76

Taylor KE, Stouffer RJ, Meehl GA (Apr 2012) An overview of CMIP5 and the experiment design. Bull Am Meteorol Soc, 485–498

Teutschbein C, Seibert J (2012) Bias correction of regional climate model simulations for hydrological climate-change impact studies: review and evaluation of different methods. J Hydrol 456–457:12–29

van Vuuren DP, Edmonds J, Kainuma M, Riahi K et al (2011) The representative concentration pathways: an overview. Clim Change 109:5–31

Weedon GP, Balsamo G, Bellouin N, Gomes S, Best MJ, Viterbo P (2014) The WFDEI meteorological forcing data set: WATCH forcing data methodology applied to ERA-Interim reanalysis data. Water Resour Res 50(9):7505–7514

Chapter 3
Population Structure and Regeneration Dynamics of Three Dominant Treeline Species from Treeline Ecotone of the Nepal Himalayas

Parveen K. Chhetri, Raju Bista, Narayan P. Gaire, and Krishna B. Shrestha

Abstract Climate change will have a strong influence on vegetation, particularly on the transition zone communities such as the treeline ecotone. The population structure and regeneration dynamics of long-lived treeline species can be utilized as an indicator of climate change and its impacts on forest vegetation. To understand the population dynamics of three dominant treeline species, *Abies spectabilis* (Himalayan Silver Fir), *Pinus wallichiana* (Blue Pine), and *Betula utilis* (Himalayan Birch) from the Nepal Himalayas, this study was conducted at the treeline ecotone of the Makalu Barun National Park, eastern Nepal; Annapurna Conservation Area, central Nepal; and Dhorpatan Hunting Reserve, western Nepal, respectively. A total of eight study plots of 20 m width and variable length, with three plots each in Makalu and Manang and two in Dhorpatan were established. We enumerated all individuals within the study transects and applied dendroecological techniques to obtain age information. The population age structure was analyzed using a static life table and survivorship curves. All three species showed the reverse J-shaped age distribution indicative of undisturbed forests with sustainable regeneration. Young individuals mostly dominated the age structure, however, their mortality was found to be very high. This could indicate limited possibilities of population densification in the near future or until the seedling mortality rate was checked. However, recruitment of all the species seems to be favored by warmer climate throughout the year, without perturbations accruing from moisture deficits. Thus, the ecotone densification could occur as temperatures continue to rise, thereby potentially decreasing mortality rates.

P. K. Chhetri (✉) · R. Bista
Department of Earth Science and Geography, California State University Dominguez Hills, Carson, CA, USA
e-mail: pchhetri@csudh.edu

N. P. Gaire
Patan Multiple Campus, Tribhuvan University, Lalitpur, Nepal
e-mail: narayan.gaire@pmc.tu.edu.np

K. B. Shrestha
Department of Biological Sciences, University of Bergen, Bergen, Norway
e-mail: krishna.shrestha@uib.no

A. Saikia and P. Thapa (eds.), *Environmental Change in South Asia*,
https://doi.org/10.1007/978-3-030-47660-1_3

Keywords Age structure · Climate change · Mortality · Population dynamics · Recruitment pattern

3.1 Introduction

Global climate is changing at an unprecedented rate and the Himalayan region is one of the hotspots of change. Climate change is impacting forest ecosystems particularly those along the treeline ecotone, the distinct upper transition zone between the closed subalpine forest and upper alpine tundra (Körner 2012). This high-altitude transition zone is mostly associated naturally with harsher climatic conditions, lower temperatures, severe winds, snow cover, and other unfavorable factors (Körner 2003; Schickhoff 2005) that limit tree survival, growth, and regeneration (Cairns and Moen 2004; Holtmeier and Broll 2007; Holtmeier 2009). These constraints make the treeline a highly stressful region for trees with a marked impact on the edge communities and even nominal changes in these conditions tend to affect the treeline ecotone. The latter can thus offer an early bellwether of forest response (Smith et al. 2009) to climate change. While the herbivory and anthropogenic disturbances also have huge influences at local to regional scale (Cairns and Moen 2004; Holtmeier and Broll 2005; Holtmeier 2009; Chhetri et al. 2017), thermal constraints are considered a primary regulator of the tree growth limitation, dieback, and seedling mortality in the unpropitious environment (Harsch and Bader 2011; Körner 2012; Crofts and Brown 2020; Holtmeier and Broll 2020). Conditions that ameliorate mechanisms constraining tree performance drive treeline responses accordingly. Thus, ongoing climate warming will likely influence vegetation communities at treeline ecotone as the warmer conditions promote tree performance in the low temperature controlled areas.

Regeneration, recruitment, and growth are essential processes shaping stand dynamics and even slight changes in stand dynamics can have considerable effects on climate-sensitive treeline ecotones. Studies have shown enhanced tree growth, survival, and better regeneration in response to climate change-induced optimal growth conditions (Harsch et al. 2012; Gaire et al. 2014; Smithers et al. 2017). However, treelines do not respond synchronously over time and scale, especially in mountain areas. In regions like the Himalayas, the growth responses of treeline trees are multidirectional, species-specific, and spatio-temporally differentiated (Schickhoff et al. 2015; Chhetri et al. 2020). There have been several instances of treeline advance and increasing recruitments at treeline ecotones across the Nepal Himalayas (Shrestha et al. 2007; Shrestha et al. 2015a; Ghimire et al. 2010; Gaire et al. 2011, 2014). However, many treelines are experiencing warming-induced growth decline in moisture deficit areas (Dawadi et al. 2013; Gaire et al. 2014, 2019; Liang et al. 2014; Tiwari et al. 2017b). In a global treeline analysis, Harsch et al. (2009) reported that only 57% of treelines experiencing temperature rise had advanced. Sharp topographic variations in mountains along with factors like disturbance, dispersal limitation, and biotic interactions result in varying responses to the climate variability (Smithers et al. 2017). The geoecological complexity and accompanying variations in hygric, thermal, edaphic, and biotic conditions bring about considerable differences in position, physiognomy, and species composition in treeline ecotone in the Himalayas (Schickhoff 2005).

Species performances such as dispersal differences, successful establishments, ecophysiological adaptational traits under changing environmental conditions, are likely determinants of regeneration patterns and stand dynamics (Malanson et al. 2007). Unprecedented change in climate patterns has an influence on speeding up community dynamics, by not only directly influencing tree growths, survival, and regeneration but also changing other biotic and abiotic controlling factors and their interactions. The relative importance of controlling factors at the treeline ecotone varies with tree stages (Körner 2012; Kulakowski et al. 2016), and thus tree survivorship and mortality under current climate change may have a differential influence on individuals. Regeneration essentials that involve seed viability, seedling establishment, survival, and growth (Körner 2012) appear challenging at the treeline ecotone (Smith et al. 2009). Survival at treeline ecotone, especially during the first year after germination, is the crucial phase of seedling establishment (Smith et al. 2009; Körner 2012; Barbeito et al. 2012; Frei et al. 2018) since the susceptibility is exacerbated by sharply fluctuating temperature and humidity (Körner 2012; Devi et al. 2020). Tree seedlings tend to be highly susceptible to drought, high radiation during the daytime, and cold nighttime temperatures (Smith et al. 2009).

The establishment of seedlings beyond the treeline ecotone indicates the upward movement of the tree species with the invasion of alpine meadows. Patterns of age distribution along the ecotone and over time archive important information on the ongoing demographic processes. Therefore, understanding the changes in historical and current structural and population dynamics across the treeline ecotone provides insights into the future direction of forest productivity and range dynamics (Dolanc et al. 2013a, b). Population stability and patterns of stand development in time and space have been assessed through age and size analyses (Ren et al. 2007; Shrestha et al. 2007; Chhetri et al. 2016). Most of the treeline studies of the Nepal Himalayas are concentrated on climate change responses (Gaire et al. 2014; Shrestha et al. 2015a; Chhetri and Cairns 2015, 2018; Sigdel et al. 2020) and few have assessed the stand structural and population dynamics emphasizing on the life table which in its simplest form contains information on the rate at which individuals from one age class survive and advance to another class (Harcombe 1987). The static life table uses the age distribution of trees in estimating the age-specific mortality and survival success, thereby quantitatively predicting the behavior of a population and depicting the direction of population dynamics (Harcombe 1987). Therefore, this study was carried out at the treeline ecotone of three sites (Makalu Barun National Park, eastern Nepal; Annapurna Conservation Area, central Nepal; and Dhorpatan Hunting Reserve, western Nepal) to understand the population dynamics of three dominant treeline species—*Abies spectabilis*, *Pinus wallichiana*, and *Betula utilis*. Our specific objectives were:

- To analyze the age structure of *Abies spectabilis*, *Pinus wallichiana,* and *Betula utilis* to assess regeneration across the treeline ecotone.
- To quantify the survivorship and mortality rate via life table to estimate the future direction of forest growth.
- To determine important climatic variables that control the recruitment pattern based on climate recruitment response analysis.

3.2 Methods

3.2.1 Study Area

In this study, we selected three study sites (Barun, Manang, and Dhorpatan) representing eastern, central, and western parts of Nepal Himalayas, respectively (Fig. 3.1), and studied the population dynamics of three treeline species—*Abies spectabilis* (Himalayan Silver Fir), *Pinus wallichiana* (Blue Pine), and *Betula utilis* (Himalayan Birch). The Barun study site was located in Barun valley of Makalu Barun National Park (27.75° N; 87.15° E; 3850–3960 m above sea level (hereafter; m asl)). Being located in the sub-tropical Asian monsoon zone, the area receives more than 80% of the annual total precipitation from June to September (Byers 1996). *Abies spectabilis* is a dominant treeline species in the area and covers south, north, and east facing slopes. *Betula utilis* (more frequent at lower elevations in the north face), *Rhododendron hodgsonii*, *R. campylocarpum*, *R. wightii*, *Acer campbelli*, and *Sorbus microphyllus* are other associated species found here (Byers 1996; Chhetri and Cairns 2015).

The Manang site was located in Ngawal of Manang valley within the Annapurna Conservation Area (28.65° N; 84.10° E; 3900–4100 m asl). Manang valley is a part of the trans-Himalayan arid region of Nepal that lies between the Tibetan plateau to the north and the main Himalayan axis to the south (Annapurna range). The valley is surrounded by many mountain ridges and steep slopes, and the climate is characterized by warm dry summers with frequent strong winds, which produce xeric conditions. Treeline ecotone ranges from 3800 to 4200 m asl and *Pinus wallichiana*, *Betula utilis*, and *A. spectabilis* are the common treeline species in the region. *Pinus wallichiana* is the dominant tree species in the study site (Ngawal), with the intermittent occurrence of *Juniperus indica* within the ecotone. Local people use the treeline ecotone area as pastures, so this area is suitable to investigate how agropastoralism interacts with the treeline ecotone dynamics, in addition to the impacts of climate change.

Study plots in Dhorpatan were located in Rugakharka (28.55° N; 83.17° E; 3850–3910 m asl) in the eastern part of Dhorpatan Hunting Reserve (hereafter DHR), western Nepal. The climate of the DHR varies from sub-tropical in the lower elevation to alpine climate in the higher elevation zone. Common flora of the reserve are

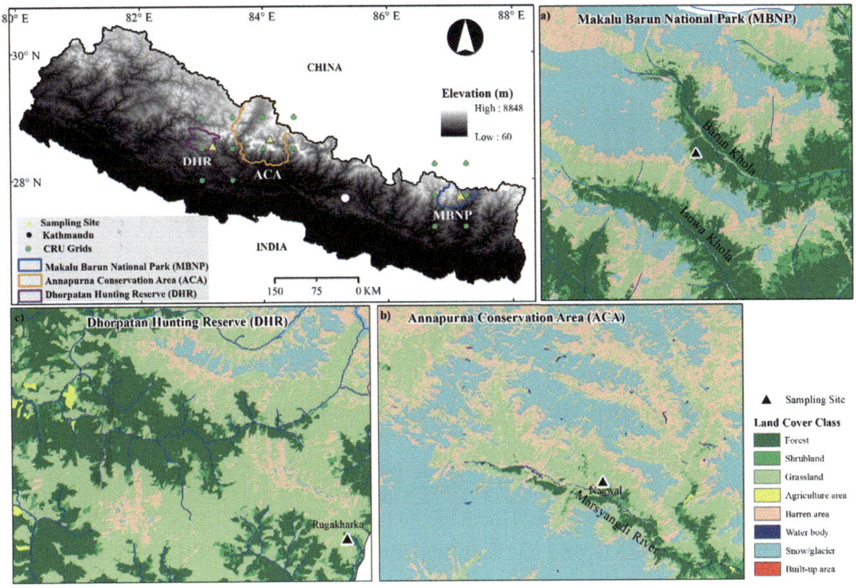

Fig. 3.1 Study sites. **a** Barun, eastern Nepal, **b** Manang, central Nepal, and **c** Dhorpatan, western Nepal (*Landcover data source* ICIMOD 2013)

Abies spectabilis, *Betula utilis*, *Pinus wallichiana*, *Quercus semecarpifolia*, *Rhododendron arboreum*, and *Rhododendron campanulatum*. *Betula utilis* is the dominant tree species in the north-facing treeline ecotone of DHR (Chhetri and Cairns 2018).

3.2.2 Study Area Climate

Eastern Nepal comparatively receives more rainfall because there is an east-to-west precipitation gradient in the central Himalaya (Shrestha et al. 2015b). Makalu Barun National Park lies within the subtropical Asian monsoon zone, and receives the majority of annual precipitation during the summer months (Byers 1996; Chhetri and Cairns 2016). The monthly average temperature is 3.5 °C and the average annual total precipitation is based on the Climate Research Unit (CRU) CRU TS v4.04 data in the grids (27.25°–27.75° N, 86.75°–87.25° E and 27.75°–28.25° N, 86.75°–87.25° E) is 1367 mm (Harris et al. 2020).

Being located in the trans-Himalayan arid region wherein high mountains of the Annapurna range act as a barrier to the south-west monsoon, Manang receives very little rainfall. The CRU grid average (28.5°–29.0° N and 84.0°–84.5° E) mean annual temperature is 5.2 °C and the annual total precipitation is 851 mm (Harris et al. 2020) which closely approximates station data (840 mm) although the latter data series are of rather limited temporal span (Shrestha et al. 2015a). The DHR study site is a mesic

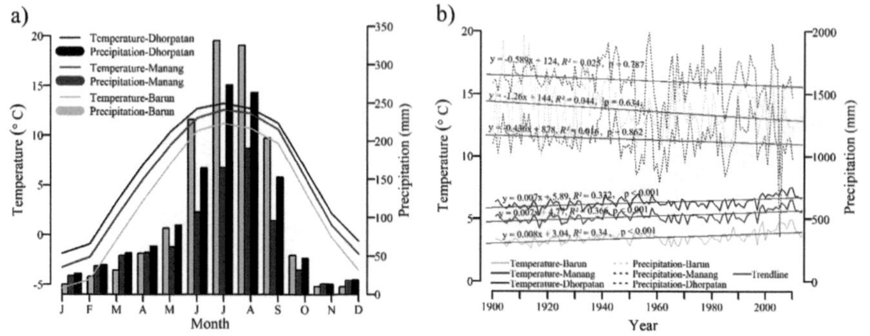

Fig. 3.2 CRU grid-based monthly (**a**) and annual (**b**) average temperature and total precipitation of three study sites in Barun, Manang, and Dhorpatan in eastern, central, and western Nepal, respectively

area with sufficient precipitation. The average CRU grid (28.0°–28.5° N, 83.0°–83.5° E and 28.5°–29.0° N, 83.0°–83.5° E) annual mean temperature of the region is 6.3 °C and the total precipitation is 1206 mm (Harris et al. 2020).

Climate stations are far away from the study sites. Moreover, those station data are of very short span and often fraught with missing data. Therefore, to acquire an overview of long-term trends, we utilzed CRU TS v4.04 datasets. We present monthly average temperature and precipitation trends of the three study sites according to long-term (1901–2016) CRU data (Harris et al. 2020) in Fig. 3.2a and b. Data does not show a clear trend in precipitation although it appears to be decreasing in all three sites. However, temperature has increased significantly in all the three sites, particularly after 1960 (Fig. 3.2b).

3.2.3 Study Species

We selected three dominant treeline species (*Abies spectabilis, Pinus wallichiana,* and *Betula utilis*), respectively, from Barun, Manang, and Dhorpatan. *Abies spectabilis* is a tall, pyramidal, evergreen tree species, which dominates the sub-alpine forests of the central Himalaya at elevations of 2700–4200 m asl in Nepal (Chhetri and Cairns 2015). Trees of this species are characterized by low branches with dense foliage. *Abies spectabilis* is the dominant treeline-forming tree in Barun with a persistent dense thicket of understory *Rhododendron hodgsonii* and *R. campylocarpum* growing profusely in the ecotone (Fig. 3.3a).

Pinus wallichiana is an evergreen conifer tree species found at an elevation of 1800–4200 m asl (Ghimire et al. 2010). This pine species is primarily confined to the dry central and western regions of Nepal. Local residents use this species as an important source of timber and fuel (Ghimire et al. 2010). *Pinus wallichiana* is

Fig. 3.3 Study species and sampling sites. **a** *A. spectabilis* from Barun. **b** *P. wallichiana* from Manang and **c** *B. utilis* from Dhorpatan

dominant at the treeline ecotone in Manang with scant *Juniperus indica* and prostrate *J. squamata* (Fig. 3.3b).

Betula utilis is a moderately sized (<20 m tall), broadleaved, pioneer tree species native to the Himalayas and is found at elevations of 2700–4500 m asl (Tiwari et al. 2017b). The species is dominant among the broadleaved tree species in the sub-alpine region of the Himalaya and is localized in moist and shady areas, where snow remains until the end of spring and sometimes even in summer (Shrestha et al. 2007). *Betula utilis* is one of the dominant species in the subalpine forest tracts of the Dhorpatan region.

3.2.4 Field Data Collection

A total of eight vertical transect plots extending from the forest line to treeline were established (Table 3.1). For this study, we defined treeline ecotone as the altitudinal zone between the forest line and tree species line (Hofgaard et al. 2009). The forest line is the upper delimitation of the closed forest and the tree species limit is the

Table 3.1 Plot characteristics of *A. spectabilis* from Barun, eastern Nepal; *P. wallichiana* Manang, central Nepal and *B. utilis* Dhorpatan, western Nepal

Characteristic parameters		Study sites		
		Barun, eastern Nepal	Manang, central Nepal	Dhorpatan, western Nepal
Study species		*Abies spectabilis*	*Pinus wallichiana*	*Betula utilis*
Number of plots		3	3	2
Plots size (m × m)		20 × 60 – 160	20 × 250	20 × 120
Aspect		North	South	North
Average slope (°)		30–35	30–35	35–40
Number of samples		184	181	186
Density (stem/ha)	Tree	101	58	141
	Sapling	90	49	39
	Seedling	455	18	206

Note Tree = Height ≥ 2 m; Sapling = Height ≥ 0.5 – <2 m; Seedling = Height < 0.5 m

uppermost elevational location of the treeline forming species. The treeline was considered as the uppermost location of erect trees with a height equal to or greater than 2 m. Three vertical transects (160 × 20 m^2) in the north-facing slope (3856–3960 m asl) in Yangle, Barun; three transects (250 × 20 m^2) across the south-facing slope (3910–4080 m asl) in Ngawal, Manang; and two transects (120 × 20 m^2) in the north-east facing slope (3850–3910 m asl) of the Rugakharka area in Dhorpatan were established. Along with the biometric measurements, trees were cored at the base, and branch whorl/bud scars were counted for age estimation of seedling/sapling individuals (Shrestha et al. 2015a; Chhetri and Cairns 2015, 2018).

3.2.5 Age Structure and Life Table Construction

We enumerated all individuals of *A. spectabilis, P. wallichiana,* and *B. utilis* within the transects and applied standard dendroecological techniques to obtain age information (Chhetri and Cairns 2015, 2018; Shrestha et al. 2015a, b). We noticed frequent multi-stemmed trees, especially in Manang. Therefore, tree cores were extracted from the base of each tree. The ages of seedlings and saplings were estimated by the branch whorls count method (Shrestha et al. 2015a). Further, the population age structure was analyzed to investigate the regeneration dynamics at the treeline ecotone. Age classes (10 years) were used for life table construction. Life tables are an important tool to estimate future survival trends of the population and can be used to investigate individual survival status in each age class at a time (Ren et al. 2007; Chhetri et al. 2016). We have presented the description of the life table components and calculation steps in Table 3.2, following Ren et al. (2007).

Table 3.2 Survival and mortality statistics of three treeline populations at different age stages

Abies spectabilis

Age class (x)	ax	lx	dx	qx	Lx	Tx	ex	lnlx	Kx
0–10	83	1000	687	0.69	657	1476	1.48	6.91	1.16
10–20	26	313	157	0.50	235	819	2.62	5.75	0.69
20–30	13	156	12	0.08	151	584	3.73	5.05	0.08
30–40	12	144	60	0.42	114	434	3.00	4.97	0.54
40–50	7	84	84	1.00	42	319	3.79	4.43	4.43
50–60	0	0	−48	0.00	24	277	0.00	0.00	0.00
60–70	4	48	0	0.00	48	253	5.25	3.88	0.00
70–80	4	48	12	0.25	42	205	4.25	3.88	0.29
80–90	3	36	−36	−1.00	54	163	4.50	3.59	−0.69
90–100	6	72	36	0.50	54	108	1.50	4.28	0.69
100–110	3	36	36	1.00	18	54	1.50	3.59	0.00
110–120	0	0	−24	0.00	12	36	0.00	0.00	0.00
120–130	2	24	12	0.50	18	24	1.00	3.18	0.69
130–140	1	12	12	1.00	6	6	0.50	2.49	2.49

Pinus wallichiana

Age class (x)	ax	lx	dx	qx	Lx	Tx	ex	lnlx	Kx
0–10	61	1000	246	0.25	877	2467	2.47	6.91	0.28
10–20	46	754	393	0.52	557	1590	2.11	6.63	0.74
20–30	22	360	66	0.18	328	1033	2.86	5.89	0.20
30–40	18	395	115	0.39	238	705	2.39	5.69	0.49
40–50	11	180	33	0.18	164	467	2.59	5.19	0.20
50–60	9	147	16	0.11	139	303	2.06	4.99	0.12
60–70	8	131	82	0.63	90	164	1.25	4.88	0.98
70–80	3	49	49	1.00	25	74	1.50	3.90	3.90
80–90	0	0	−16	0.00	8	49	0.00	0.00	0.00
90–100	1	16	−16	0.00	25	41	0.00	0.00	−3.49
100–110	2	32	33	1.00	16	16	0.50	3.49	3.49
110–120									
120–130									
130–140									

(continued)

Table 3.2 (continued)

Betula utilis									
Age class (x)	ax	lx	dx	qx	Lx	Tx	ex	lnlx	Kx
0–10	47	1000	43	0.04	979	3457	3.46	6.91	0.04
10–20	45	957	319	0.33	798	2479	2.59	6.86	0.41
20–30	30	638	277	0.43	500	1681	2.63	6.46	0.57
30–40	17	362	213	0.59	500	1181	3.26	5.89	0.89
40–50	7	149	−43	−0.29	255	926	6.21	5.00	−0.25
50–60	9	191	−128	−0.67	170	755	3.94	5.25	−0.51
60–70	15	319	191	0.60	223	500	1.57	5.77	0.92
70–80	6	128	64	0.50	96	277	2.17	4.85	0.69
80–90	3	64	−0	0.00	64	181	2.83	4.16	0.00
90–100	3	64	21	0.33	53	117	1.83	4.16	0.41
100–110	2	43	21	0.50	32	64	1.50	3.75	0.69
110–120	1	21	0	0.00	21	32	1.50	3.06	0.00
120–130	1	21	21	1.00	11	11	0.50	3.06	3.06
130–140									

Ax: survival individuals at the starting age class; lx: proportion of individuals surviving from the beginning to age x, supposing 1000 individuals of the first age class (lx = ax/ao); dx: number of dead individuals from age x to x + 1 (dx = ax − ao); qx: individual mortality from x to x + 1 (qx = dx/lx); Lx: mean number of survival individuals from age x to x + 1 (Lx = (lx + lx + 1)/2); Tx: total number of survival individuals from age x (Tx = ΣLx); ex: life expectancy at age x (ex = Tx/lx); Kx: population mortality rate, Kx = ln(lx) − ln(lx + 1). Calculating formulae from Ren et al. (2007), Chhetri et al. (2016)

Based on the life table of *A. spectabilis*, *P. wallichiana,* and *B. utilis* population, the survivorship and mortality curves were calculated. The survivorship curve was drawn between the lnlx (proportion of individuals surviving from the beginning age group to that particular age group) and the age class, while the mortality and killing curve were drawn between qx (mortality from the age group x to the next x + 1 age group), Kx (population mortality rate) and the age class (Harcombe 1987).

3.2.6 *Recruitment and Climate Response*

To determine the principal climatic factors limiting establishment patterns at the treeline ecotone, a relationship between climate and recruitment pattern was established. We used CRU data (Harris et al. 2020) for response analysis. Climate variables (monthly mean temperature and monthly total precipitation) of the current year were used for the climate recruitment response analysis. In addition, the seasonal relationship of Spring (March, April, May), Summer (June, July, August), Autumn (September, October, November), and Winter (December, January, February) mean temperature and total precipitation with recruitment number was also investigated. In this study, climate data was compiled into 10-year averages. Similarly, recruitment periods of established individuals were grouped into 10-year interval classes. Climate and recruitment data were compared using Pearson's correlation. This information would be helpful in predicting the future response of sub-alpine treeline species to climate change (Jump et al. 2007).

3.3 Results

3.3.1 Age Structure

All three treeline species viz. *A. spectabilis* from Barun, eastern Nepal, *P. wallichiana* Manang, central Nepal, and *B. utilis* Dhorpatan, western Nepal showed reverse J-shape age structure curves (Fig. 3.4). In all three study sites, stand histories marginally exceeded a century and higher proportions of individuals tended to belong to younger age classes, as did the density distribution (Table 3.1). The percentage of young individuals (less than 10 years) were higher for *A. spectabilis* in Barun than two other study species—*P. wallichiana* (Manang) and *B. utilis* (Dhorpatan). However, no significant differences in the number of individuals (less than 100 years) were discernible in all three species. Thus, age distribution showed successful and sustained recruitment over time in all species. However, there appeared a slight halt in the *A. spectabilis* establishment in Barun during the 1950s. In Manang treeline as well, only three of the oldest trees were established in the first two decades of the twentieth century and regeneration gradually increased by mid-century with a considerable surge in regeneration by the 1970s.

3.3.2 Survivorship and Mortality Curves

Survivorship curves of *A. spectabilis, P. wallichiana,* and *B. utilis* are almost Deevey-III type with some peculiarities in the curves of the *B. utilis* and *P. wallichiana* (Fig. 3.5). The survivorship of the species gradually declined with age, with a sharp plunge for the *B. utilis* at 50–60 years and in *P. wallichiana* at 80–100 years, possibly indicative of local site-specific adverse events for the corresponding periods. Other than this, all species showed similar rates overall.

The static life table (Table 3.2) and mortality curves (Fig. 3.6) showed that young cohorts have a high mortality rate. The mortality of *A. spectabilis* (individual of below 10 years) was 69%, that of *P. wallichiana* was 25%, and *B. utilis* mortality at age below 10 years was only 4%. There was a slight difference in mortality in the subsequent age group. The 60–70 year group of *A. spectabilis* had the highest life expectancy and that of the *B. utilis* was obvious for the 40–50 year cohort. Mortality

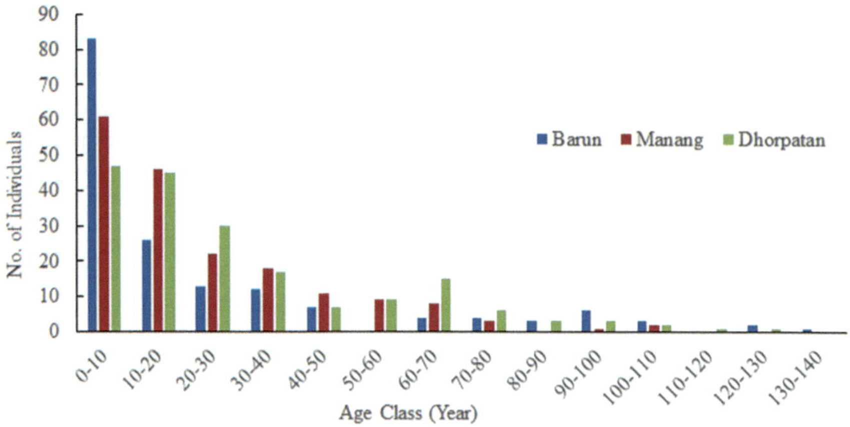

Fig. 3.4 10-year age structure of *A. spectabilis* from Barun, *P. wallichiana* from Manang and *B. utilis* from Dhorpatan in eastern, central, and western Nepal, respectively

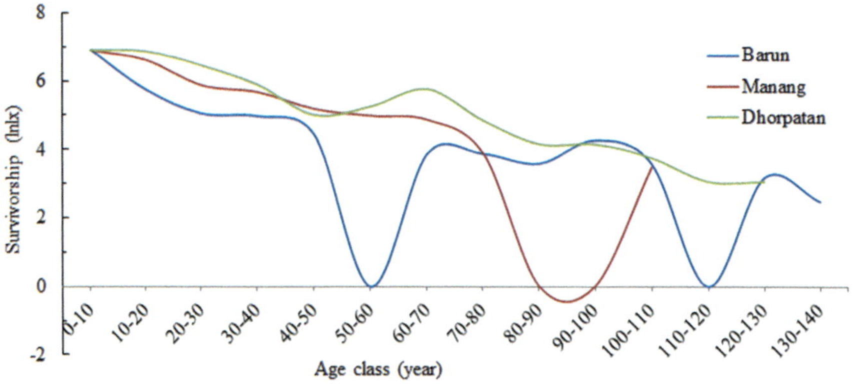

Fig. 3.5 Survivorship curves of *A. spectabilis* from Barun, eastern Nepal, *P. wallichiana* Manang, central Nepal and *B. utilis* Dhorpatan, western Nepal. (lx: proportion of individuals surviving from the beginning to the age x ($lx = a_x/a_o$, where, a_0 is the individual number in starting age group and a_x is the individual number in the x age class)

rate curves and killing power (hazard rate) curves of all three species showed similar fluctuations (Fig. 3.6). The first mortality peak for *A. spectabilis* was at the starting age class, and then the bizarre peak occurred at 40–50 years. The first considerable mortality phase of *P. wallichiana* was at 10–20 years with an uncharacteristic peak (at 70–80) and depression (at 90–100 years). *Betula utilis* mortality appeared to peak at 30–40 years of age.

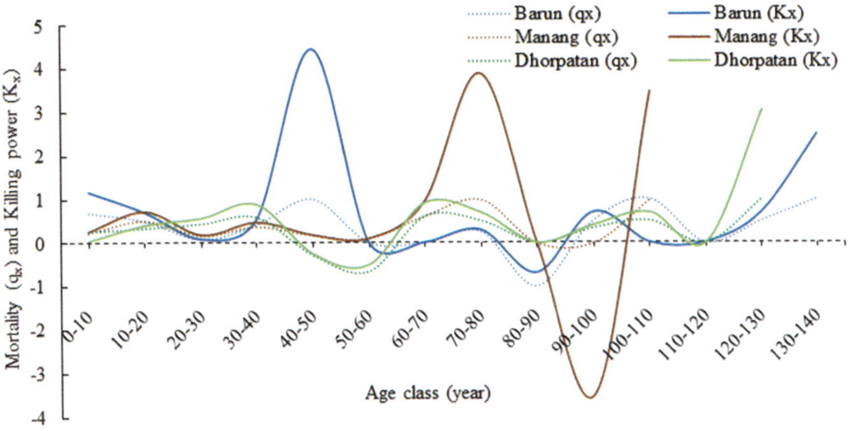

Fig. 3.6 Variation trend of mortality (q_x) and killing power (K_x) *A. spectabilis* (Barun, eastern Nepal), *P. wallichiana* (Manang, central Nepal), and *B. utilis* (Dhorpatan, western Nepal)

3.3.3 Recruitment and Climate Response

Recruitment patterns showed a higher affinity to temperature (monthly and seasonal) than the precipitation for all the species (Fig. 3.7). *Abies spectabilis* (in Barun) showed significant positive correlation with spring (MAM) ($r = 0.73$, $p < 0.05$), summer (JJA) ($r = 0.75$, $p < 0.05$), autumn (SON) ($r = 0.78$, $p < 0.05$), and winter (DJF) ($r = 0.87$, $p < 0.01$) temperature. Significant positive correlation was also observed with spring (MAM) ($r = 0.67$, $p < 0.05$) precipitation for this species. For the same species, the current monthly temperature showed a significant positive correlation except for May and June. May precipitation also showed significant positive correlation ($r = 0.85$, $p < 0.05$). *Pinus wallichiana* in Manang showed a significant positive correlation to the monthly mean temperature of all months excluding May, June, and July, where the species had a significant negative correlation with August precipitation. Similarly, *B. utilis* establishment showed a significant positive correlation with the temperature of all months except May, June, July, and September, but exhibited a negative correlation with precipitation during the summer months.

3.4 Discussion

The population dynamics and regeneration pattern of treeline species in the Nepal Himalayas are likely to change in response to climate change (Gaire et al. 2014, 2019; Tiwari et al. 2017a). Our analyses provide the understandings of forest stand structure and regeneration dynamics at three treeline ecotones of the Nepal Himalayas. Our results showed that irrespective of the population of different species and variations in transitional distances (Table 3.1), in different regions, those treeline stands are

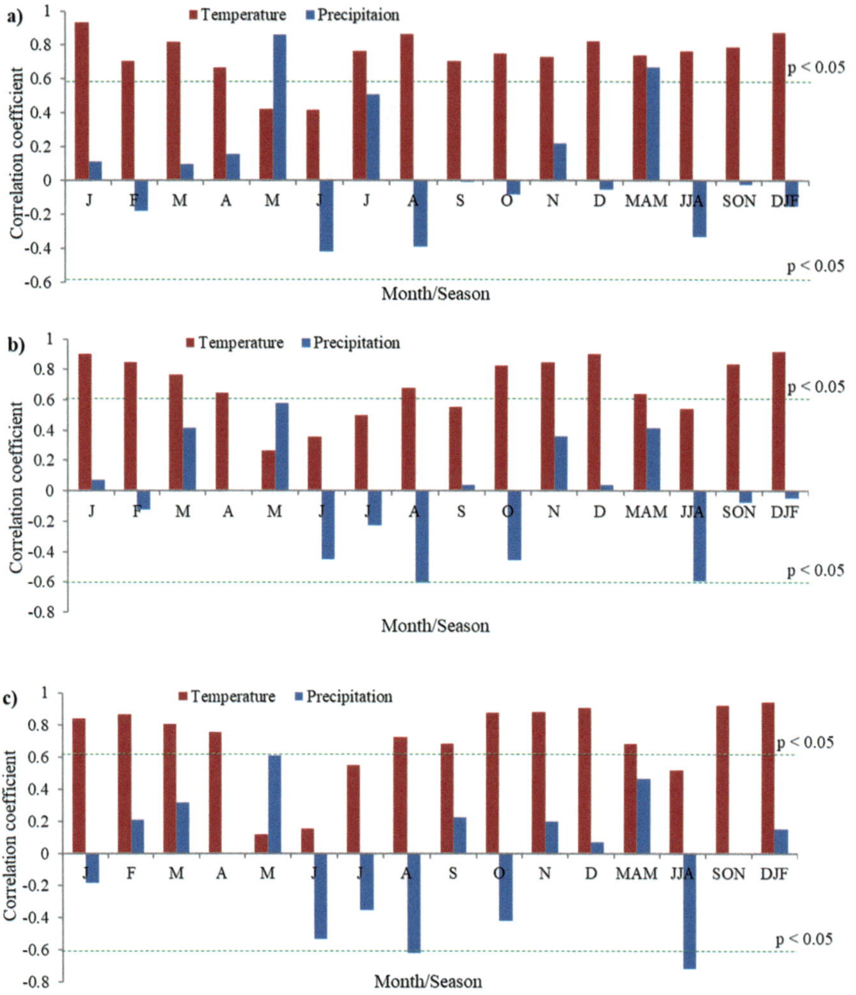

Fig. 3.7 Recruitment pattern and climate correlation analysis. **a** *A. spectabilis* from Barun, eastern Nepal. **b** *P. wallichiana* from Manang, central Nepal, and **c** *B. utilis* from Dhorpatan, western Nepal. Months are indicated by the initial letters, MAM (Spring), JJA (Summer), SON (Autumn), and DJF (Winter)

almost in equilibrium, where sufficient seedlings and saplings stipulate the dynamics. All three ecotonal stands did not show considerable gaps in their age distribution. Therefore, we can argue that these populations are self-maintaining. In a forest, large disturbances leading to high tree mortality, or episodes of poor regeneration, thereby create gaps in stand structure. Therefore, the instances of atypical plunges in survivorship curves in some age classes, and similar uncharacteristic peaks or dips in mortality curves may indicate the influence of other adverse non-climatic disturbance factors. Otherwise, overall trends in those curves, recruitment responses to climate, and occurrences of high frequency younger cohorts may imply a satisfactory natural regeneration of these species in the high altitude edge environment of the Himalayas.

3.4.1 Population Dynamics

Successful seedling establishment, growth, and survival are the important prerequisites for treeline advance and treeline ecotone densification (Kambo and Danby 2018), and the life history traits, climatic variations, and various disturbance regimes influence the processes. Understanding population dynamics in a transitional zone like treeline is more important when/where a slight climate change can produce a considerable change in the community processes. Age structure and static life table can provide a fair picture of population dynamics over time (Harcombe 1987; Ren et al. 2007). The reverse-J-shaped age structure of all species showed significant recruitment in the last two or three decades. The young stand history in all the ecotones and the regeneration surge after mid latter half of the twentieth century indicate favorable conditions in the treeline ecotone either due to climatic amelioration or land-use changes. Rai et al. (2013) observed very poor regeneration of *A. spectabilis* compared to *B. utilis* in Western Himalaya India, with a low seedling-to-sapling conversion ratio. However, several studies on the *A. spectabilis* treeline (Ghimire and Lekhak 2007; Gaire et al. 2014, 2017; Chhetri and Cairns 2016), the *P. wallichiana* treeline (Ghimire et al. 2010; Shrestha et al. 2015a), and the *B. utilis* treeline (Shrestha et al. 2007; Gaire et al. 2017) are a testament of ecotonal densification in recent decades in different parts of the Nepal Himalayas.

Life tables show the age-cohort-specific survival and mortality rate which are the important population characteristics that drive population processes. Deevey (1947) proposed three types of mortality (survivorship) curves in which the logarithm of the proportion of a first cohort surviving to a given age is plotted against age. A convex Deevey-I curve indicates increased death risk with age in the population. A type II declining line shows a constant risk over all ages. A concave type III curve is indicative of decreasing death risk with age, which is the most common type of curve for all organisms (Harcombe 1987). Fluctuating survivorship curves indicate the heterogeneous influence of climate and environmental conditions over time(Ren et al. 2007), in the study sites. Mortality peak before 20 years of age is related to high mortality in seedlings (more than 50% in *A. spectabilis* and *P. wallichiana,* and just above 30% in *B. utilis*). In juvenile mortality, *A. spectabilis* seedlings had the highest likelihood (69% mortality of seedlings less than 10 years). *Abies spectabilis* grows on moist shady slopes and is sensitive to late spring frosts (Shrestha et al. 2015a). In the rather open ecotone, lack of moist substrates like moss coverage, shades, and light limitation may cause hardships for seedlings to thrive. The mortality peak at the second age cohort of *P. wallichiana* in Manang, may not be indicative of that age group higher mortality propensity. We estimate that the recent increase of human influence in the ecotone (Ghimire and Lekhak 2007; Shrestha et al. 2015a) may have facilitated the regeneration surge during the last two decades, with comparatively low numbers of the cohorts established immediately prior to land-use change. Another possible reason could be the role of nurse plants. There were considerable numbers of seedlings and saplings in the dense thickets of *Juniperus indica*. Nurse plants provide a better micro environment through ameliorated soil and temperature. When the seedlings emerge, exposure to winter frost or snow, wind abrasion, and other stressors can make them vulnerable (Holtmeier 2009; Körner 2012). This may be how seedling mortality, instead of initial stages, delayed the peak. In the *P. wallichiana* population, three of those trees aged around 100 years were from the forest line. The absence of trees established before the 1930s was plausibly the reason underlying the abnormal peak and depression in the mortality curves of the rather young population. Despite the relatively shorter transition, the number of small *B. utilis* individuals in Dhorpatan may indicate ameliorating conditions in the ecotone and the mortality peaking only at the age of thirties.

Static life table and mortality curves, however, do not always show population processes precisely (Ren et al. 2007). During the life history of long-lived species, climatic variations and disturbances have differential influences over time. The static life table assumes static influences of environmental conditions (Ren et al. 2007). For example, as we see on the high peak on the mortality curve of 40–50 years for the *A. spectabilis*, it does not necessarily imply that the age cohort is vulnerable. Instead, the specific case may be because of disturbances of any kind, or a very unfavorable climate leading to the recruitment obstacle in the period or high population mortality in 1950s. The decade was not climatically unfavorable compared to before and after the decade (Fig. 3.2b). Non-climatic factors most likely might have suppressed the regeneration. The removal of a significant number of trees or stymied recruitment due to such an extreme event can raise a biased generalization that under the normal conditions too, that age cohort most likely has low survivorship.

3.4.2 Regeneration and Climate

Climate and recruitment response analysis can help assess climatic variables associated with high mortality in younger age classes in the treeline ecotone (Chhetri et al. 2016). Reproduction and mortality both depend on individual tree growth performances, and climatic amelioration promotes establishment at treeline (Jump et al. 2007). Therefore, higher survival or lower mortality rate can imply the alleviated environmental stress in the ecotone. The lower temperature in the ecotone limits growth as well as seedling establishments. The positive relationships between recruitment and summer, autumn, and winter season temperatures could be related to a longer growing season and reduced mortality due to warmer conditions (higher daily minimum temperatures, fewer frosts) (Camarero and Gutierrez 1999). However, a weak relationship was found between precipitation and the recruitment pattern of these regions. Wang et al. (2012) in southeastern Tibetan Plateau, and Zhao et al.

(2013) in Min Mountain, southwestern China also found no relationship between the recruitment number and precipitation pattern at treeline. These highlight the positive effects of a warming climate in the temperature-controlled ecotone, and precipitation not yet has been short.

While regional warming may have a general trend of influence on treeline vegetation, the prevailing local circumstances control the directionality of population and community dynamics. Moreover, species-specific traits of the response also are not similar. For example, Gaire et al. (2014, 2017) reported spatio-temporal variability in recruitment, with a high rate of ecotone densification and the advance of *A. spectabilis* in Manasalu and Sagarmatha. Recruitment in those treelines correlated with temperature all around the year, similar to the response from Barun. Lv and Zhang (2012) also found that recruitment of an *A. spectabilis* in the Tibetan side of Mt. Everest is sensitive to temperature. Chhetri and Cairns (2016) found moisture limitation insignificant in the radial growth of the species in Barun, although the growth was significantly correlated with winter temperature (similar response in Humla, Gaire et al. 2020). Therefore, slightly below average precipitation without any unusual temperatures by the early 1950s (Fig. 3.2) alone may not explain the moisture constraints in the *A. spectabilis* recruitment break of 1950s. We can expect an episodic regeneration at treeline (Lv and Zhang 2012; Gaire et al. 2014), and thus patchy distribution in size or age classes can be characteristic of this shade-tolerant species (Gaire et al. 2014). The *A. spectabilis* seedlings in Barun, more often than not, were found under the canopy. In comparison, seed dispersal distance was not unduly high; whereas the mortality of small individuals was high as could be expected.

Shrestha et al. (2015b) found a weak positive radial growth response of *P. wallichiana* to summer temperature and negative growth response to winter precipitation. The negative influence of winter precipitation was attributed to its role in increased snow accumulation, late spring snow thawing, and shorter growing seasons. However, the recruitment was correlated to the monthly mean temperature of all months barring May, June, and July. The recruitment relationship with summer precipitation was negative. Seedlings are generally light-demanding (Ghimire et al. 2010). The high summer precipitation implies high cloud cover, reduced insolation, and lower soil temperature (Shrestha et al. 2015b), and even the chance of snow accumulation. These adversities hinder seedling establishment. Similar was the recruitment response of *B. utilis* in Dhorpatan. *Betula utilis* is a shade-intolerant tree species that depend on more gaps for regeneration (Taylor and Zisheng 1992; Shrestha et al. 2007). Several experimental studies have shown poor recruitment in the forest understory and more seedling survival with better light availability from the gap created from self-thinning or disturbances (Shrestha et al. 2007). A warming experiment by Xu et al. (2012) showed that warming favored *B. utilis* at high altitudes through the extended growing season, accelerated leaf/shoot functional traits, and enhanced photosynthetic capacity. Being a deciduous species, inter-annual climatic variations are more important in *B. utilis* phenology, and spring phenophases are sensitive to late winter and early spring temperature (Rai et al. 2013). Warm temperature is positively correlated to its bud break. However, warming-induced moisture deficit has a negative influence on its radial growth (Dawadi et al. 2013; Shrestha et al. 2017; Tiwari et al. 2017a). Therefore, climate response is spatio-temporal. *Betula utilis* in the Manasalu and Sagarmatha were spatio-temporally inconsistent in terms of regeneration responses, with low regeneration in stable treelines (Gaire et al. 2014, 2017). Chhetri and Cairns (2018) also observed a low recruitment of *B. utilis* at the Dhorpatan treeline. In contrast, there was a sufficient regeneration of *B. utilis* in Manang (Shrestha et al. 2007). In Dhorpatan, however, considering short ecotonal stretches, regeneration appeared to be self-sustaining, and warming in the moist area could support the regeneration continuation, provided the herbivory influence was reduced.

An almost similar consistency in mortality of the three species does not necessarily show similar regeneration success. Important bottlenecks for seedlings and saplings to emerge as trees persist. A shade-intolerant species may have a high mortality rate. Moreover, herbivory may be the barrier (Chhetri and Cairns 2018). Therefore, even if there were a sufficient number of young cohorts, the grazing pressure, shades below treeline, and predisposition to drier conditions in upper open areas could be reasons behind high juvenile mortality rates. The age distribution, survivorship curves, and recruitment favorability of warmer climate show that all three treeline stands have a healthy regeneration.

3.5 Conclusions

We applied a static life table and survivorship curve approach to understanding the population and regeneration dynamics of three dominant treeline species (*A. spectabilis* from Barun, eastern Nepal, *P. wallichiana* Manang, central Nepal, and *B. utilis* Dhorpatan, western Nepal) from the Nepal Himalayas. An almost reverse J-shaped age structure indicated a progressive regeneration and densification in the ecotone. However, a high mortality rate in the youngest age classes may indicate the impediment in the process. Since higher mortality in initial stages and subsequent reduced risk with age (type III survivorship curve) are common in nature, juvenile mortality can be expected. Despite being in different parts of the Nepal Himalayas, and with slightly different climate settings, all species' recruitment was found to be favored by warming. Therefore, by casting aside site-specific disturbances, warming seems to be enhancing ecotone densification.

References

Barbeito I, Dawes MA, Rixen C, Senn J, Bebi P (2012) Factors driving mortality and growth at treeline: a 30-year experiment of 92 000 conifers. Ecology 93(2):389–401. https://doi.org/10.1890/11-0384.1

Byers AC (1996) Historical and contemporary human disturbance in the upper Barun valley, Makalu-Barun National Park and Conservation Area, east Nepal. Mt Res Dev 16(3):235–247. https://doi.org/10.2307/3673946

Cairns DM, Moen J (2004) Herbivory influences tree lines. J Ecol 92:1019–1024. https://doi.org/10.1111/j.1365-2745.2004.00945.x

Camarero JJ, Gutierrez E (1999) Structure and recent recruitment at alpine forest pasture ecotones in the Spanish central Pyrenees. Ecoscience 6:451–464. https://doi.org/10.1080/11956860.1999.11682540

Chhetri PK, Cairns DM (2015) Contemporary and historic population structure of *Abies spectabilis* at treeline in Barun valley, eastern Nepal Himalaya. J Mt Sci 12(3):558–570. https://doi.org/10.1007/s11629-015-3454-5

Chhetri PK, Cairns DM (2016) Dendroclimatic response of *Abies spectabilis* at treeline ecotone of Barun Valley, eastern Nepal Himalaya. J For Res 27(5):1163–1170. https://doi.org/10.1007/s11676-016-0249-7

Chhetri PK, Bista R, Cairns DM (2016) Population structure and dynamics of *Abies spectabilis* at treeline ecotone of Barun Valley, Makalu Barun National Park, Nepal. Acta Ecol Sin 36(4):269–274. https://doi.org/10.1016/j.chnaes.2016.05.003

Chhetri PK, Shrestha KB, Cairns DM (2017) Topography and human disturbances are major controlling factors in treeline pattern at Barun and Manang area in the Nepal Himalaya. J Mt Sci 14(1):119–127. https://doi.org/10.1007/s11629-016-4198-6

Chhetri PK, Cairns DM (2018) Low recruitment above treeline indicates treeline stability under changing climate in Dhorpatan Hunting Reserve, western Nepal. Phys Geogr 39(4):329–342. https://doi.org/10.1080/02723646.2018.1428266

Chhetri PK, Bista R, Shrestha KB (2020) How does the stand structure of treeline-forming species shape the treeline ecotone in different regions of the Nepal Himalayas? J Mt Sci 17(10):2354–2368. https://doi.org/10.1007/s11629-020-6147-7

Crofts AL, Brown CD (2020) The importance of biotic filtering on boreal conifer recruitment at alpine treeline. Ecography 43(6):914–929. https://doi.org/10.1111/ecog.04899

Dawadi B, Liang E, Tian L, Devkota LP, Yao T (2013) Pre-monsoon precipitation signal in tree rings of timberline *Betula utilis* in the central Himalayas. Quat Int 283:72–77. https://doi.org/10.1016/j.quaint.2012.05.039

Deevey ES Jr (1947) Life tables for natural populations of animals. Q Rev Biol 22(4):283–314

Devi NM, Kukarskih VV, Galimova AA, Mazepa VS, Grigoriev AA (2020) Climate change evidence in tree growth and stand productivity at the upper treeline ecotone in the Polar Ural Mountains. For Ecosyst 7(1):7. https://doi.org/10.1186/s40663-020-0216-9

Dolanc CR, Thorne JH, Safford HD (2013a) Widespread shifts in the demographic structure of subalpine forests in the Sierra Nevada, California, 1934 to 2007: shifting structure of subalpine forests in California. Glob Ecol Biogeogr 22:264–276. https://doi.org/10.1111/j.1466-8238.2011.00748.x

Dolanc CR, Westfall RD, Safford HD, Thorne JH, Schwartz MW (2013b) Growth–climate relationships for six subalpine tree species in a Mediterranean climate. Can J For Res 43:1114–1126. https://doi.org/10.1139/cjfr-2013-0196

Frei ER, Bianchi E, Bernareggi G, Bebi P, Dawes MA, Brown CD, Trant AJ, Mamet SD, Rixen C (2018) Biotic and abiotic drivers of tree seedling recruitment across an alpine treeline ecotone. Sci Rep 8:10894. https://doi.org/10.1038/s41598-018-28808-w

Gaire NP, Dhakal YR, Lekhak HC, Bhuju DR, Shah SK (2011) Dynamics of *Abies spectabilis* in relation to climate change at the treeline ecotone in Langtang National Park. Nepal J Sci Technol 12:220–229. https://doi.org/10.3126/njst.v12i0.6506

Gaire NP, Dhakal YR, Shah SK et al (2019) Drought (scPDSI) reconstruction of trans-Himalayan region of central Himalaya using *Pinus wallichiana* tree-rings. Palaeogeogr Palaeoclimatol Palaeoecol 514:251–264. https://doi.org/10.1016/j.palaeo.2018.10.026

Gaire NP, Fan ZX, Shah SK, Thapa UK, Rokaya MB (2020) Tree-ring record of winter temperature from Humla, Karnali, in central Himalaya: a 229 years-long perspective for recent warming trend. Geogr Ann A 102(3):297–316. https://doi.org/10.1080/04353676.2020.1751446

Gaire NP, Koirala M, Bhuju DR, Borgaonkar HP (2014) Treeline dynamics with climate change at the central Nepal Himalaya. Clim Past 10(4):1277–1290. https://doi.org/10.5194/cp-10-1277-2014

Gaire NP, Koirala M, Bhuju DR, Carrer M (2017) Site-and species-specific treeline responses to climatic variability in eastern Nepal Himalaya. Dendrochronologia 41:44–56. https://doi.org/10.1016/j.dendro.2016.03.001

Ghimire BK, Lekhak HD (2007) Regeneration of *Abies spectabilis* (D. Don) Mirb. in subalpine forest of upper Manang, north-central Nepal. In: Chaudhary RP, Aase TH, Vetaas OR, Subedi BP (eds) Local effects of global changes in the Himalayas: Manang, Nepal, Tribhuvan University, Nepal and University of Bergen, Norway, pp 139–149

Ghimire B, Mainali KP, Lekhak HD, Chaudhary RP, Ghimeray AK (2010) Regeneration of *Pinus wallichiana* AB Jackson in a trans-Himalayan dry valley of north-central Nepal. Himal J Sci 6(8):19–26. https://doi.org/10.3126/hjs.v6i8.1798

Harcombe PA (1987) Tree life tables. Bioscience 37(8):557–568. https://doi.org/10.2307/1310666

Harris I, Osborn TJ, Jones P et al (2020) Version 4 of the CRU TS monthly high-resolution gridded multivariate climate dataset. Sci Data 7:109. https://doi.org/10.1038/s41597-020-0453-3

Harsch MA, Bader MY (2011) Treeline form – a potential key to understanding treeline dynamics. Glob Ecol Biogeogr 20(4):582–596. https://doi.org/10.1111/j.1466-8238.2010.00622.x

Harsch MA, Buxton R, Duncan RP, Hulme PE, Wardle P, Wilmshurst J (2012) Causes of tree line stability: stem growth, recruitment and mortality rates over 15 years at New Zealand *Nothofagus* tree lines. J Biogeogr 39(11):2061–2071. https://doi.org/10.1111/j.1365-2699.2012.02763.x

Harsch MA, Hulme PE, McGlone MS, Duncan RP (2009) Are treelines advancing? A global meta-analysis of treeline response to climate warming. Ecol Lett 12(10):1040–1049. https://doi.org/10.1111/j.1461-0248.2009.01355.x

Hofgaard A, Dalen L, Hytteborn H (2009) Tree recruitment above the treeline and potential for climate-driven treeline change. J Veg Sci 20(6):1133–1144. https://doi.org/10.1111/j.1654-1103.2009.01114.x

Holtmeier FK (2009) Mountain timberlines: ecology, patchiness, and dynamics, 2nd edn. Springer, Germany

Holtmeier F-K, Broll G (2005) Sensitivity and response of northern hemisphere altitudinal and polar treelines to environmental change at landscape and local scales. Glob Ecol Biogeogr 14(5):395–410. https://doi.org/10.1111/j.1466-822X.2005.00168.x

Holtmeier F-K, Broll G (2007) Treeline advance–driving processes and adverse factors. Landsc Online 1:1–33. https://doi.org/10.3097/LO.200701

Holtmeier FK, Broll G (2020) Treeline research—from the roots of the past to present time. A review. Forests 11(1):38. https://doi.org/10.3390/f11010038

ICIMOD (2013) Land cover of Nepal 2010. ICIMOD. https://doi.org/10.26066/rds.9224

Jump AS, Hunt JM, Peñuelas J (2007) Climate relationships of growth and establishment across the altitudinal range of Fagus sylvatica in the Montseny Mountains, northeast Spain. Ecoscience 14(4):507–518. https://doi.org/10.2980/1195-6860(2007)14[507:CROGAE]2.0.CO;2

Kambo D, Danby RK (2018) Factors influencing the establishment and growth of tree seedlings at Subarctic alpine treelines. Ecosphere 9(4):e02176. https://doi.org/10.1002/ecs2.2176

Körner C (2003) Alpine plant life: functional plant ecology of high mountain ecosystems, 2nd edn. Springer, New York

Körner C (2012) Alpine treelines: functional ecology of the global high elevation tree limits. Springer, Basel, p 220. https://doi.org/10.1007/978-3-0348-0396-0

Kulakowski D, Barbeito I, Casteller A, Kaczka RJ, Bebi P (2016) Not only climate: interacting drivers of treeline change in Europe. Geogr Pol 89(1):7–15. https://doi.org/10.7163/GPol.0042

Liang EY, Dawadi B, Pederson N, Eckstein D (2014) Is the growth of birch at the upper timberline in the Himalayas limited by moisture or by temperature? Ecology 95:2453–2465. https://doi.org/10.1890/13-1904.1

Lv LX, Zhang QB (2012) Asynchronous recruitment history of *Abies spectabilis* along an altitudinal gradient in the Mt Everest Region. J Plant Ecol 5(2):147–156. https://doi.org/10.1093/jpe/rtr016

Malanson GP, Butler DR, Fagre DB, Walsh SJ, Tomback DF, Daniels LD, Resler LM, Smith WK, Weiss DJ, Peterson DL, Bunn AG, Hiemstra CA, Liptzin D, Bourgeron PS, Shen Z, Millar CI (2007) Alpine treeline of western North America: linking organism-to-landscape dynamics. Phys Geogr 28(5):378–396. https://doi.org/10.2747/0272-3646.28.5.378[notcitedintext]

Rai ID, Bharti R, Adhikari BS, Rawat GS (2013) Structure and functioning of timberline vegetation in the western Himalaya: a case study. In: Wu N, Rawat GS, Joshi S, Ismail M, Sharma E (eds) High-altitude rangelands and their interfaces in the Hindu Kush Himalayas. ICIMOD, Kathmandu, pp 91–107

Ren Q, Yang X, Cui G, Wang J, Huang Y, Wei X, Li Q (2007) Smith fir population structure and dynamics in the timberline ecotone of the Sejila Mountain, Tibet, China. Acta Ecol Sin 27(7):2669–2677. https://doi.org/10.1016/S1872-2032(07)60055-9

Schickhoff U (2005) The upper timberline in the Himalayas, Hindu Kush and Karakorum: a review of geographical and ecological aspects. In: Broll G, Keplin B (ed) Mountain ecosystems. Springer, Berlin, Heidelberg, pp 275–354. https://doi.org/10.1007/3-540-27365-4_12

Schickhoff U, Bobrowski M, Böhner J et al (2015) Do Himalayan treelines respond to recent climate change? An evaluation of sensitivity indicators. Earth Syst Dyn 6:245–265. https://doi.org/10.5194/esd-6-245-2015

Shrestha BB, Ghimire B, Lekhak HD, Jha PK (2007) Regeneration of treeline birch (*Betula utilis* D. Don) forest in a trans-Himalayan dry valley in Central Nepal. Mt Res Dev 27:259–267. https://doi.org/10.1659/mrdd.0784

Shrestha KB, Chhetri PK, Bista R (2017) Growth responses of *Abies spectabilis* to climate variations along an elevational gradient in Langtang National Park in the central Himalaya, Nepal. J For Res 22(5):274–281. https://doi.org/10.1080/13416979.2017.1351508[notcitedintext]

Shrestha KB, Hofgaard A, Vandvik V (2015a) Recent treeline dynamics are similar between dry and mesic areas of Nepal, central Himalaya. J Plant Ecol 8(4):347–358. https://doi.org/10.1093/jpe/rtu035

Shrestha KB, Hofgaard A, Vandvik V (2015b) Tree-growth response to climatic variability in two climatically contrasting treeline ecotone areas, central Himalaya, Nepal. Can J for Res 45(11):1643–1653. https://doi.org/10.1139/cjfr-2015-0089

Sigdel SR, Liang E, Wang Y, Dawadi B, Camarero JJ (2020) Tree-to-tree interactions slow down Himalayan treeline shifts as inferred from tree spatial patterns. J Biogeogr 47(8):1816–1826. https://doi.org/10.1111/jbi.13840

Smith WK, Germino MJ, Johnson DM, Reinhardt K (2009) The altitude of alpine treeline: a bellwether of climate change effects. Bot Rev 75(2):163–190. https://doi.org/10.1007/s12229-009-9030-3

Smithers BV, North MP, Millar CI, Latimer AM (2017) Leap frog in slow motion: divergent responses of tree species and life stages to climatic warming in Great Basin subalpine forests. Glob Chang Biol 24(2):e442–e457. https://doi.org/10.1111/gcb.13881

Taylor AH, Zisheng Q (1992) Tree regeneration after bamboo die-back in Chinese Abies-Betula forests. J Veg Sci 3(2):253–260. https://doi.org/10.2307/3235687

Tiwari A, Fan ZX, Jump AS, Li SF, Zhou ZK (2017a) Gradual expansion of moisture sensitive *Abies spectabilis* forest in the Trans-Himalayan zone of central Nepal associated with climate change. Dendrochronologia 41:34–43. https://doi.org/10.1016/j.dendro.2016.01.006

Tiwari A, Fan ZX, Jump AS, Zhou ZK (2017b) Warming induced growth decline of Himalayan birch at its lower range edge in a semi-arid region of Trans-Himalaya, central Nepal. Plant Ecol 218(5):621–633. https://doi.org/10.1007/s11258-017-0716-z

Wang Y, Camarero JJ, Luo T, Liang E (2012) Spatial patterns of Smith fir alpine treelines on the south-eastern Tibetan Plateau support that contingent local conditions drive recent treeline patterns. Plant Ecol Divers 5:311–321. https://doi.org/10.1080/17550874.2012.704647

Xu Z, Hu T, Zhang Y (2012) Effects of experimental warming on phenology, growth and gas exchange of treeline birch (*Betula utilis*) saplings, Eastern Tibetan Plateau, China. Eur J For Res 131(3):811–819. https://doi.org/10.1007/s10342-011-0554-9

Zhao ZJ, Shen GZ, Tan LY, Kang DW, Wang MJ, Kang W, Guo WX, Zeppel MJB, Yu Q, Li JQ (2013) Treeline dynamics in response to climate change in the Min Mountains, southwestern China. Bot Stud 54(1):1–12. https://doi.org/10.1186/1999-3110-54-15

Chapter 4
Environmental Changes and Rural Livelihoods in Muktapur Village, Kamrup District, Assam, India

Haruhisa Asada, Yusuke Yamane, and Nityananda Deka

Abstract Northeast India is a region endowed with diverse nature and culture. It is also one of the regions in the country where the economically backward population make a living from subsistence agriculture which largely depends on the local environment. In Assam, about 75% of the total 31 million population is engaged in agriculture and related activities, which suggests that the livelihood of the state is highly dependent on nature. Since the local natural environment is closely related to the livelihoods of villagers, it is important to investigate the changes in the environment at a micro-scale. Natural environment in rural villages has hardly been studied so far, however, because it is difficult to analyze the detailed features from satellite images, and secondary data on micro-environments is not available. This study will reveal the changes in the natural environment in a rural village in Assam from the perspective of the locals. Characteristics and changes of the agroecosystem, transformation of recent agricultural land use, and perceptions of villagers about climate changes are examined in the study village.

Keywords Agroecosystem · Land-use change · Climate change · Rural livelihoods · Assam

4.1 Introduction

Northeast India is a region endowed with diverse nature and culture. It is also one of the regions in the country where economic development has remained tardy, and

H. Asada (✉)
Department of History, Sociology and Geography, Nara Women's University, Nara, Japan
e-mail: asada@cc.nara-wu.ac.jp

Y. Yamane
Faculty of Education, Tokoha University, Shizuoka, Japan
e-mail: yamane@sz.tokoha-u.ac.jp

N. Deka
Department of Geography, Gauhati University, Guwahati, India
e-mail: nityananda@gauhati.ac.in

A. Saikia and P. Thapa (eds.), *Environmental Change in South Asia*,
https://doi.org/10.1007/978-3-030-47660-1_4

agriculture remains the predominant livelihood. Assam is sandwiched in the middle, with alluvial lowlands formed by the Brahmaputra river and abundant rainfall brought by the southwest monsoon that provides suitable conditions for rice cultivation. About 75% of its population of some 31 million is engaged in agriculture and related activities (Government of Assam 2017), which suggests that the livelihoods of the state are highly dependent on nature.

Most geographical studies on the natural environment of Assam have been carried out from macro-scale as well as meso-scale perspectives, and few have dealt with the natural environment on a micro-scale. For example, studies on the macro scale revealed the characteristics of the ecological zones of the Brahmaputra valley (e.g. Gopalkrishnan 2000; Bhagabati et al. 2001), and the extent of flood hazards (e.g. Kar 2014), and studies on the meso-scale examined changes in forest coverage (e.g. Saikia 2013) and resource of wetlands and their surrounding environment (e.g. Baruah 2018) using remote sensing and GIS. On the other hand, the natural environment in rural villages has hardly been studied thus far because it is difficult to analyze the detailed features from satellite images and secondary data is not always readily available. However, since the local natural environment is closely related to the livelihoods of villagers, it is important to investigate the changes of the environment on a micro-scale.

In order to reveal the natural environment of villages and its changes, it is effective for researchers not only to conduct field surveys but also to cooperate with the local inhabitants. Villagers are usually familiar with the local natural environment, and the influence of the natural environment often appears on their livelihood patterns, especially, on farming. Therefore, it will be possible to estimate the changes in the natural environment by listening to the experiences and indigenous knowledge of villagers and investigating changes in their livelihood activities. In previous studies, the role of local inhabitants was almost completely ignored in geographical analyses of the natural environment, but it is necessary to reveal the changes in the natural environment from the perspective of the local residents.

This study examines the environmental changes in the Brahmaputra valley of Assam in general and its micro-environment and the changes therein in the study of a sample village in particular. Since rural environments are constantly changed and shaped by the interaction between human activities and nature, this study focuses on agriculture, which is the main livelihood of villagers, and the climate which affects agriculture and agricultural land use. We will reveal that how changes in agriculture and climate lead to dynamic changes in the micro-level environment in the village.

4.2 Study Area

Field survey was carried out in Muktapur village in Kamrup district, Assam. The village is located on the north bank floodplain of the Brahmaputra river, and is about 35 km from Guwahati city towards the north and 40 km south of the Bhutan Himalayan foothills. This is a typical village inhabited by indigenous non-tribal

Assamese people. There are 491 households with a population of 2,068, covering an area of 3.67 km^2 (household survey of 2017). The village consists of 11 hamlets, each with a different caste composition. The 11 hamlets were formed in different years, but it is estimated that the old hamlet was established some 250 years ago and the new about 150 years ago (Bhagabati and Deka 2016).

Results in the following sections are derived from micro-level observations in the village and interviews with villagers. In Sect. 4.3, characteristics and changes of the agroecosystem in the village are explained, and its implications on villagers' livelihoods are discussed based on the long-term observation of the third author (Deka). In Sect. 4.4, transformation of recent agricultural land use and its social background are described based on the observation and interview of the first author (Asada). In Sect. 4.5, perceptions of villagers about climate changes are examined based on the interview of the second author (Yamane). Finally, in the conclusion section, environmental changes in the village and its relationship with rural livelihoods and society are discussed.

4.3 Agroecosystem and Livelihood Issues

4.3.1 Characteristics of Agroecosystem

The livelihood patterns of Muktapur village are basically characterized by foodgrain and cash crop farming, livestock rearing, fisheries, traditional homestead gardening, kitchen gardening etc., which form the basis of sustainable livelihood of the villagers. The land-use patterns and practices are basically governed by the village's micro-ecological settings on the one hand and the culture and traditions of the inhabitants on the other.

The traditional farming system, although relatively less productive, is more often than not eco-friendly and sustainable (Deka and Bhagabati 2010). One of the most important advantages of traditional agroecosystems is that farmers usually get an opportunity to acquire a good understanding of the concerned ecosystems working within the limits of the area, which may have great potentiality to contribute to the long-term sustainability of the local agroecosystems (Fig. 4.1). Therefore, the traditional agroecosystem of the village has immense ecological, economic and social significance.

As far as the protection of crops from insects and pests is concerned, farmers used age-old practices and organic substances. Farmers used wood-ash on crops, such as chili, onion, potato, brinjal, lady's finger, cucumber, etc. Another common practice for controlling diseases and insects was to apply the leaves or branches of certain medicinal plants, such as pasatia (Vitex negundo) and mahaneem (Azadirachta Indica) on the paddy fields. Farmers also tended to hang rotten crabs on the paddy fields so that its smell could attract insects and thus deviate them from the targeted crops (Fig. 4.2).

Fig. 4.1 Traditional agricultural practices

Fig. 4.2 Traditional methods of crop protection and manuring

The traditional agroecosystems of Muktapur village, like other parts of the Brahmaputra valley are bio-physically very diverse as they comprise numerous varieties of domesticated and wild plants and animals. The raising of home gardens, grazing lands, kitchen gardens, fisheries, forestry, wet rice cultivation, rotational fallowing and the traditional horticulture and cash crop farming systems contribute to the rich agroecosystem of the village (Deka 2012). The integrated crop-livestock-fish farming system, which is also a typical character of the village agroecosystem, contributes towards sustainable yields and biological pest control depending upon the internal resources and recycling of nutrients and organic matter and trophic relationship among plants, insects or pathogens.

4.3.2 Changes in Agroecosystem and Their Implications

The agroecosystem of Muktapur village, like other parts of the Brahmaputra valley, has been changing both qualitatively and quantitatively over time due to the nature and degree of modification of the agricultural systems made by the local dwellers (Deka 2012). With the rapid growth of population, changes in socio-economic conditions, growing food requirements, introduction of modern agricultural technologies and implementation of some non-judicious Government plans and policies, the diversity and long-term sustainability of the village's traditional agroecosystem have witnessed perceptible changes (Deka et al. 2017).

After the 1980s, the farmers of Muktapur village started using certain modern agricultural inputs and implements. The farmers of the village are usually ignorant about the detrimental effects of chemical fertilizers and pesticides on the ecosystem and human health. Although there is enough scope for modern irrigation systems in the village, this is currently lacking as a result of which they are unable to cultivate crops all throughout the year. Because of erratic monsoonal rain, low market price and limited market facilities for organic agricultural products and low productivity of traditional agriculture, the farmers of the village were compelled to practice modern agriculture. However, since the farmers of the village are economically poor and marginal, they could not afford to procure expensive agricultural inputs and implements.

Previously almost all the farmers had a pair of bullocks and other indigenous agricultural tools to cultivate local varieties of crops. They used these tools based on the available natural resources without incurring any expenses. These tools had also immense ecological significance since they caused no harm to the farm- friendly insects and also maintain natural soil fertility. The farmers also shared agricultural tools amongst themselves based on mutual understanding and cooperation.

However, due to the growing influence of innovative technology on agriculture, some age-old cropping technologies are gradually disappearing in the village. To cultivate HYV (High Yielding Variety) crops, farmers have to depend on power tillers and tractors as they have to be cultivated within a stipulated time span. In Muktapur village only two or three households own power tillers and tractors for which the rest of the farmers have to depend on to till their lands, and also they have to pay money which is a constraint for many residents. Another important point to be noted here is that the agricultural lands of the village are fragmented into very small sizes and scattered over space (Table 4.1), rendering it inconvenient in implementing modern agricultural practices.

Thus, agriculture had become a costly enterprise for the poor and marginal farmers of the village. Thus, many farmers have, at present, left agricultural activities as a result of which many problems have emerged in the village, such as farm abandonment and transformation of farm lands to other land uses, occupational shift, rural depopulation etc. The disappearance of traditional agricultural practices has caused adverse impacts not only on the agroecosystem but also on the food habit, house types, culture and the very sustainability of the rural economy.

Table 4.1 Area under different landholding size-classes

Size class (ha)	Total land in hectare	No. of households possessing land
Below 1	117.6 (40.8)	321(79)
1–2	84.4 (29.3)	60 (15)
2–3	42.0 (14.6)	17 (4)
3–4	21.2 (7.4)	6 (1)
Above 4	23.2 (8.1)	4 (1)
Total	288.4	408

Source Field survey, 2006–2007
Note Figures in the parentheses indicate the percentage to the total

Another disheartening fact is that the farmers of the Brahmaputra valley are not well-trained under any Government scheme regarding the use of chemical fertilizers and pesticides on their agricultural fields. Therefore, they tend to unscientifically apply chemical inputs on their agricultural fields causing great degradation of soil health. Again, as the agricultural fields of the village are closely connected to wetlands, therefore, the chemicals applied on their fields ultimately drain into the neighboring wetlands during the rainy season harming aquatic floral and faunal species. At present, many fish and other aquatic species have declined in the wetlands leading to the breaching of food-chain processes in the village ecosystem. Very recently, it has been noticed that due to the disappearance of certain predatory birds with the loss of aquatic and amphibian species, the population of monkeys and squirrels has increased drastically. The latter have severely damaged the rural economy by destroying betel nuts, coconuts and other fruits and vegetables.

There has been a gradual increase in the proportion of the lands ploughed by power tillers, while the lands ploughed by bullock-driven wooden plough have been gradually decreasing in the study village. The agricultural lands in the village have been converted to HYV rice crops at an increasing pace. The area under HYV rice in the village increased by 20.7% during the period of 2001–2007 while the area under traditional rice varieties decreased by 21.9%. During the same period, the proportion of area with fertilizer input increased by a not insubstantial 42.4%. However, in the case of application of chemical pesticides it shows a negative change of 20.0% (Table 4.2).

It is important to note that although the production of the local rice variety is low in comparison to HYV rice, the former provides multiple benefits to farmers. The market value of local rice varieties is very high, and they can be used for preparing many traditional snacks like rice cakes (locally called Pitha), and their straw can be used for roof making. Many local rice plants are flood and insect resilient and they usually do not need the application of chemical fertilizers and pesticides. Again the winter rice Sali varieties are used in many rituals and ceremonies.

The farmers of Muktapur village used to grow different varieties of crops in order to meet their basic needs (Table 4.3). In the early days, as a majority of the people of the village were unable to meet the expense of purchasing basic goods, therefore, the agricultural lands were put to double or triple cropping. It is found that the area

Table 4.2 Changes in area under the use of agricultural inputs and implements

Year	Land ploughed (in ha)		Area under rice (in ha)		Area under fertilizer use (in ha)		Area under pesticides use (in ha)	
	By wooden plough	By power tillers	Local	HYV	Organic	Chemical	Organic	Chemical
2001	127.3	100.3	108.5	91.4	109.3	80.6	58.5	25.4
2007	94.2	118.2	84.7	110.3	98.3	114.7	41.7	20.3
Change (%)	−26.0	+17.9	−21.9	+20.7	−10.1	+42.4	−28.7	−20.0

Source Field survey, 2007
Note Data for 2001 were obtained through oral interviews with elderly farmers

Table 4.3 Change in area under different crops

Year	Area (in ha) under different crops						
	Rice	Oil seeds	Pulses	Sugarcane	Potato	Vegetables	Jute
2001	199.88	3.66	3.99	2.37	9.79	5.33	2.53
2007	195.02	6.69	4.36	0.13	4.62	2.12	0.32
Change (%)	−2.43	+82.79	+9.27	−94.51	−52.81	−60.23	−87.35

Source Field survey and oral interview among the elderly farmers, 2007
Note Data for 2001 were obtained through oral interviews with elderly farmers

under rice has been gradually decreasing in the village. The decline of rice acreage is primarily due to the monocropping of HYV rice, shrinkage of cropped lands and transformation of paddy lands to other non-agricultural uses.

4.4 Transformation of Agriculture and Land-Use Change

4.4.1 Agricultural Trends in Assam

Recently, labor shortages in the agricultural sector have emerged as a challenge in Muktapur village due to rural–urban migration. The growth of the urban industrial sector has attracted rural workers into the labor force. Since Muktapur is not far from the urban center, Guwahati city, an increasing number of villagers have sought white-collar jobs or entrepreneurial pursuits rather than continuing to work in agriculture, a sector which requires considerable input costs and hard labor. Thus, interest in agriculture has gradually declined among the village's youth. Faced with decreasing rural populations and declining interest in agriculture, farming activities cannot be sustained at the former levels, leading to farm abandonment and associated environmental changes.

Several additional factors have contributed to this trend. The first is low rice yields in recent years. In the financial year (FY) 2016, rice yields averaged 2.1 t/ha in Assam, which were lower than the national average of 2.6 t/ha and only half that in the highest yielding state of Punjab (4.0 t/ha) (Government of India 2017). Along with Bihar and Odisha, Assam's rice yields are among the lowest in the country. Low yields are related to the prevalence of HYVs. The ratio of HYVs to total rice cultivation (in terms of acreage) was 71% in FY2014 and only 66% for Sali rice, which is the main rainy-season rice crop (Government of Assam 2016). While HYVs accounted for 95% for dry-season Boro rice cultivation, only 10% of the cultivated land is irrigated in Assam; thus, areas cultivated with Boro rice represented only approximately 14% of the total cultivated land. Without adequate irrigation facilities, HYVs, and chemical fertilizers, rice yields in Assam remain much lower than those in other Indian states, making it challenging for farmers to earn sufficient income via rice cultivation.

The second factor is unstable weather conditions. Monsoon rainfall is characterized by large inter-annual variations and has presented an overall decreasing trend after 2000 (Deka et al. 2013). When precipitation rates are low in June and July, the first half of the rainy season, transplanting work cannot be performed at an appropriate time on non-irrigated cultivated land, which greatly affects yields (Asada 2012).

The third factor is increasing wages for agricultural laborers. Rice cultivation requires wage labor for transplantation and other tasks; however, in Assam, workers' daily wages have increased by approximately four times over the past 10 years (from Rs 65 per day in 2006 to Rs 241 per day in 2015). Thus, landowners struggle to hire laborers for rice cultivation because the costs exceed the benefits. Farm owners represented 51% of the total workforce in Assam in 1991; however, this figure decreased to 36% in 2011 (Government of Assam 2017). Over the same period, however, the percentage of agricultural laborers within the workforce has decreased only slightly from 12 to 10%. These trends may reflect a situation in which the income is better for wage laborers than for farmers who employ laborers. While the proportion of people engaged in agriculture is on the decline, statistics show that the proportion of other (non-farmer) workers increased from 36 to 51% from 1991 to 2011 (Government of Assam 2017).

The fourth factor involves the economic policies of the Indian government. The Food Corporation of India (FCI) regularly purchases more rice and wheat produced in Punjab and Haryana than in Assam. In 2016, the FCI purchased 372,443.12 t of rice from Assam, which is only approximately 7% of Assam's total rice production of 5,127,000 t (Government of Assam 2017). Furthermore, India's public distribution system (PDS) discourages local farmers from growing rice in their fields. PDS distributes rations, including rice produced in Punjab and Haryana, to 'above poverty line' (APL), 'below poverty line' (BPL), and 'Antyodaya Anna Yojana' ('poorest of the poor', AAY) households at prices far below the market level. In villages in Assam, households that do not have non-farm-related income usually fall into one of these categories, and thus can access the rice they eat on a daily basis at a reasonably low price. Since government policy does not encourage local farmers in Assam to practice subsistence agriculture, Assam's rural residents have little incentive to grow staple crops in their own fields.

These factors have combined to de-incentivize farming in Assam, leading to farm abandonment. Several effects of farm abandonment have begun to appear, including the decreased extent of cultivated land, changes in paddy field ecosystems and changes in rural landscapes. The resultant changes in land use that have occurred as a result of farm abandonment are discussed in the context of Muktapur village. Both cultural and economic factors influencing current land-use patterns are discussed.

4.4.2 Conversion of Farmland to Forest Plots

Much of the cultivated land has been converted to allow for uses other than rice cultivation. In the 2018 kharif season, 6.7 ha of agricultural land in Muktapur village was not cultivated and remained fallow due to lack of water or labor or because of reduced productivity, especially at higher elevations. Apart from fallow lands, abandoned cultivated lands are mainly used as either forest lands or fishponds according to their size and distance from the owners' residences.

In Muktapur, 63 of a total of 491 households converted some part of their cultivated land to forests (Fig. 4.3). An average of 0.04 ha per household was converted in this manner, representing approximately 11% of each household's cultivated land. This trend has been observed since the 1990s; previously, useful trees that were sometimes planted within residential areas were planted in cultivated land where crops were grown. Since the mid-1990s, 1–5 village households planted trees within cultivated land annually.

Cultivated land planted with trees generally have certain specific characteristics. The first is its relatively high elevation. The village is located in the central part of the Brahmaputra floodplain, in which natural levees and backwater swamps are formed and the topography is characterized by a gentle relief with a specific height of only a few meters. Paddy fields at relatively high elevations are used as seedbeds and paddy fields for Ahu rice grown in the early periods of the rainy season, but the topographical and soil conditions discourage water retention, leading to low rice productivity. Therefore, villagers tend to abandon rice cultivation on higher lands and plant trees instead. Secondly, cultivated land located at a distance from residential areas is often selected for tree planting. The land owned by villagers is often fragmented inside and outside the village, and it is challenging for villagers to visit cultivated land located several kilometers away from their residences on a daily basis (Deka et al. 2011). In addition, in cases where distant land is worked by sharecroppers, landowners may not be able to monitor their tenants, who may misreport yields and thus seek to reduce the rent payable to the landowner. The reduced economic benefits associated with rice cultivation make landowners more likely to convert cultivated land to other uses.

Planting trees on cultivated land allows owners to sell timber after the trees have grown. Gomari (Gmelina arborea), Chegun (Techtona grandis), Jori(Ficus hispida), and Kodom(Anthocephaluscadamba) are the most popular tree species grown in

Fig. 4.3 Trees planted in paddy fields

cultivated lands in Muktapur village. Except for Jori, which is mainly used as firewood, these tree species all yield valuable timber. In particular, Gomari and Chegun are known to produce high-quality timber and can be sold for more than Rs 20,000 each; however, these species require about 20–25 years of growth to reach a sufficient size for sale. Landowners must purchase tree seedlings from a nearby market and water and fertilize the seedlings with cow dung after transplanting; however, thereafter, the trees require little maintenance and can rely on rainwater. However, several years of growth is required before the trees can be sold, and during growth, it remains unclear whether the expected income will be obtained.

Why are trees planted in previously cultivated lands if the profits from selling timber may not be high? The villagers widely hold that they should not simply leave cultivated lands idle. If lands are left uncultivated for no specific reason, villagers feel they may be judged by other households or that the lands may be illegally usurped by others before the landowner knows it. Thus, especially when cultivated land is located far away from the owner's residence, crops may be planted arbitrarily on the land if owners have not clearly indicated their possession and use of the land. Thus, the planting of trees prevents others from using and occupying land, granting owners peace of mind.

4.4.3 Conversion of Farmland to Fisheries

Along with tree planting, villagers in Muktapur have also been engaged in the creation of fish ponds on once-cultivated land. Of the 491 households in the village, 164 converted some portion of their cultivated lands into fishponds; thus, this approach is more common than the conversion of cultivated land to forest land. The average area of fishponds is 0.08 ha, which is equivalent to approximately 20.0% of the total cultivated land owned by a household. Fishponds are less noticeable than tree stands, making it more difficult to observe such changes in the rural landscape. However, the permanence of this conversion strongly suggests that the villagers' interest in engaging in agriculture is steadily diminishing.

The construction of fishponds is not a new practice in Muktapur. Traditionally, when sons become independent of their parents and construct residences at new sites, soil is typically sourced from cultivated lands, and the site of the excavation is often used to create a pond (Deka and Bhagabati 2015). Such ponds are called Kal and do not have levees. During rainy season floods, wild fish can freely enter it from surrounding waters. Fish species caught in Kal include Puti (Puntius chola), Goroi (Channa punctatus), Magur (Clariasmagur), and Singi (Heteropneustes fossilis). These fish are used exclusively for household consumption and cannot be sold for income.

Since the 2000s, however, ponds have been dug in cultivated land for different purposes. Some households started to convert cultivated lands previously used for crop production into ponds to raise fish for income. Such ponds are called Pukuri, and they are separated from the surrounding ecological environment by 1 to 1.5 m banks on all sides (Fig. 4.4) (Deka and Bhagabati 2015). Rather than collecting wild fish, Pukuris are used for aquaculture through the introduction of fry. The most cultivated fish species include Rou (Labeo rohita), Mirika (Cirrhinusmrigala), Bakuwa (Gibelioncatla), and Silver Carp (Hypophthalmichthys molitrix). These fish fetch high prices on the market, and some can be sold for more than Rs 500 kg.

Cultivated land that is converted into fishponds typically presents some common features. First, cultivated land that is relatively low-lying is selected. Some low-lying cultivated lands are inundated during the rainy season, making drainage difficult. As such land is not suitable for rice cultivation, it is readily converted into fishponds. In addition, land plots of a certain size tend to be selected for conversion to ponds. In some cases, an entire plot of cultivated land is converted into a fish pond, and in others, rice is cultivated in the remaining parts. For households that own several cultivated plots in the village, land near the residence tends to be selected for conversion into fishponds. This is because the fish may be stolen from the pond at night, and owners can monitor fishponds close to their residences to prevent theft.

Farming households convert cultivated land into fishponds due to economic factors since fish farming requires less labor than rice cropping and offers relatively larger profits. Households that experience a loss in labor due to the deaths of husbands or sons sometimes cannot maintain their paddy fields and thus convert them into ponds. Unlike rice cultivation, in which income is unstable, income obtained from

Fig. 4.4 Fishery ponds in paddy fields

fish raising is stable and relatively high. In households where sons do not hold regular jobs, fathers sometimes build fishponds to secure stable incomes for their sons that will inherit the land.

However, the construction of fishponds can be challenging and costly for individual farmers. A landowner first contacts a construction agent called Thikedar. During the rice growing period in rainy-season, Thikedars bring tractors to cultivate the farmland for sowing, but during the dry season, there is no demand for paddy cultivation; therefore, these individuals are typically engaged in road construction and construction of residential projects. Thikedars use heavy machinery to excavate the cultivated land, and if the landowner wishes to retain the excavated soil, he will have to bear the cost for excavation. However, if the landowner does not need it, the contractor can purchase it on the spot, often bringing the excavation fee close to zero. Through this arrangement, cultivated land can be excavated and converted into a fishpond even by households with limited means.

For landowners, the construction of fishponds is more advantageous than conversion to forest land not only in terms of financial costs but also in terms of labor costs. To sell fish raised in the pond at a profit, the fish must be caught and transported to a market. The task of catching fish in the pond is more labor intensive than harvesting rice. Hindus in the village may catch fish for personal consumption but rarely offer them for sale for religious reasons.

However, Muslims living in neighboring villages play an important role in the conversion of cultivated land to fishponds and in bringing the fish to market. Muslim merchants not only catch fish, but also sell eggs and feed to help Hindu landowners begin farming fish. It is said that in the past, Muslim earth diggers (Mati Kata) would manually dig ponds in paddy fields rather than using machines. Often, Muslim fish merchants (Mach Bepari) living in the surrounding villages are invited to catch fish from ponds owned by Hindus, and a group of 5 to 6 merchants currently live in the area around Muktapur village. When summoned, they use a huge net that covers the entire pond to catch the fish, and the landowners are paid on the spot according to the weight of the fish caught. Thus, landowners can earn income from fishponds with minimal effort.

The Muslims fish merchants tend to have a large number of male children; thus, each son inherits relatively little farmland from their parents and cannot support their families via farming alone. Catching and selling fish around Hindu villages allows them to earn an income. Previously, they caught fish only during the dry season when the Kals dried up; however, as the number of Pukuri ponds in cultivated land has increased, these Muslim fishermen visit more villages by bicycle with their nets all year around.

Presently, the role of agricultural land is changing in Muktapur village. In the past, village residents used their land to produce staple foods, but now residents attempt to earn income to meet their economic needs. Changes in villagers' mindsets have led to new land uses, which affect the surrounding environment at a micro-level.

4.5 Climate Change, Extreme Weather and People's Perception

4.5.1 Trends of Daily Maximum and Daily Minimum Temperature

Changes of the agroecosystem and agricultural land-use pattern can be partly attributed to the recent climate change. As most of the villagers are directly or indirectly associated with rice-based agriculture which is solely dependent on rainwater without any irrigation facilities such as tube wells, climate and its changes have a great impact on the socio-economic system in the village.

Figure 4.5 shows the yearly variations of daily maximum and minimum temperature at the grid closest to Muktapur village from 2000 to 2019 for annual, pre-monsoon (March to May), monsoon (June to September), post-monsoon (October to November), and winter (December to February). The annual daily maximum temperature exhibits an increasing trend from 2000 to 2019. The rate of increase is defined as the slope of linear regression line that is 0.4° per 10 years. The daily maximum temperature during the pre-monsoon, monsoon, post-monsoon, and winter also shows an increasing trend. The rates of these increases are 0.3° per 10 years, 0.6° per 10 years, 0.4° per 10 years, and 0.2° per 10 years, respectively. The rate of increase in daily maximum temperature during the monsoon is the largest.

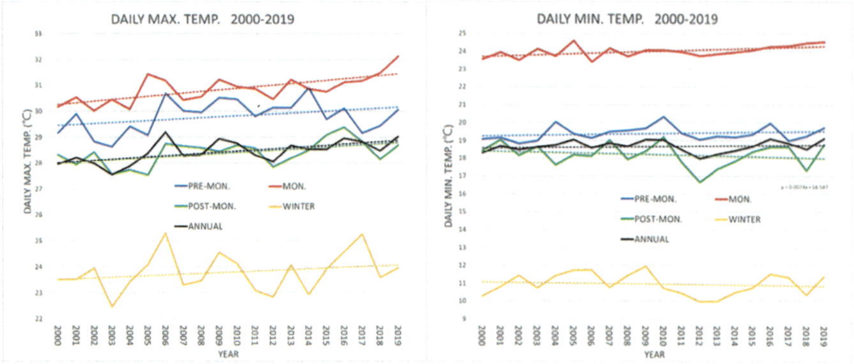

Fig. 4.5 Yearly variations of daily maximum and minimum temperature for annual, pre-monsoon, monsoon, post-monsoon, and winter from 2000 to 2019. *Source* Grid data provided by the India Meteorological Department (IMD). http://www.imdpune.gov.in/Clim_Pred_LRF_New/Grided_Data_Download.html. Data of the grid closest to Muktapur village (91.5° E, 26.5° N for daily maximum and daily minimum temperature, and 91.75° E, 26.5°N for daily rainfall) were used

The annual daily minimum temperature shows increasing trend from 2000 to 2019. However, the rate of increase is 0.07° per 10 years and is extremely negligible. The rate of change for the pre-monsoon, monsoon, post-monsoon, and winter is 0.1° per 10 years, 0.3° per 10 years, −0.2° per 10 years, and −0.1° per 10 years. Increasing trends can be found in the pre-monsoon and monsoon, and the rate of increasing is greater during the monsoon. On the contrary, decreasing trends can be found in the post-monsoon and winter. In the post-monsoon and winter, the daily maximum temperature and the daily minimum temperature tends to decrease. This may indicate that the daily difference between maximum and minimum temperature has become larger year after year in the post-monsoon and winter seasons.

4.5.2 Trends of Daily Rainfall

Figure 4.6 shows the yearly variations of rainfall at the grid closest to Muktapur village from 2000 to 2019 for annual, pre-monsoon, monsoon, post-monsoon, and winter. Annual precipitation tends to decrease from 2000 to 2019. The rate of decreasing is −600 mm per decade. Decreasing trends can be found in the pre-monsoon, monsoon, post-monsoon, and winter seasons, with the rate of changing of at −140 mm, −357 mm, −99 mm, and 9.8 mm per decade, respectively. In particular, it is characteristic that rainfall in the monsoon tends to decrease with a higher rate of change.

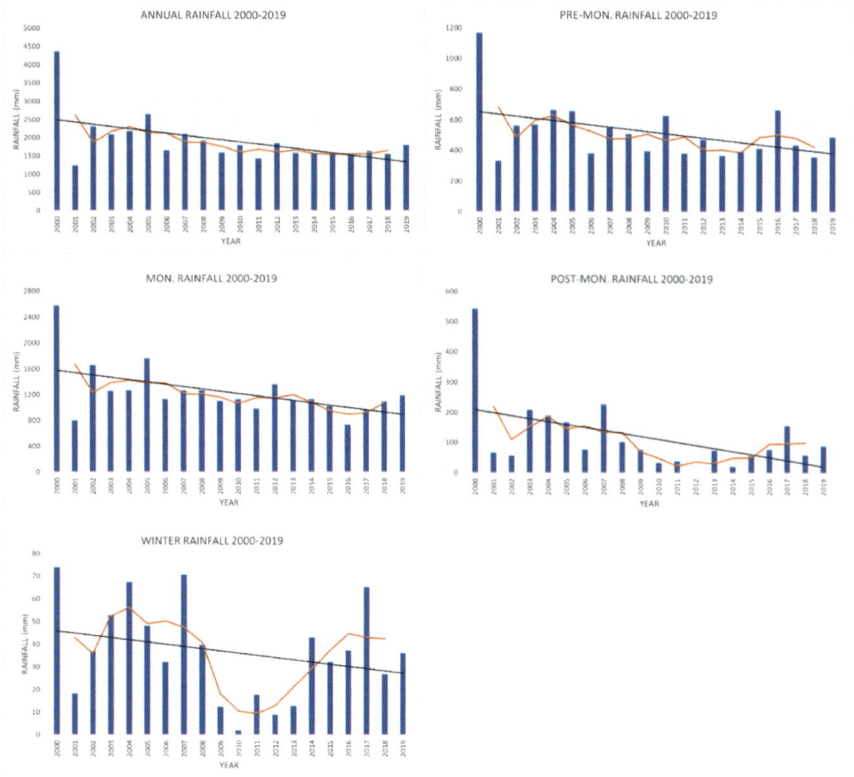

Fig. 4.6 Yearly variations of rainfall for annual, pre-monsoon, monsoon, post-monsoon and winter from 2000 to 2019. *Source* Grid data provided by the India Meteorological Department (IMD). http://www.imdpune.gov.in/Clim_Pred_LRF_New/Grided_Data_Download.html. Data of the grid closest to Muktapur village (91.5° E, 26.5° N for daily maximum and daily minimum temperature, and 91.75° E, 26.5°N for daily rainfall) were used

4.5.3 Perceptions of Local People for Validity of Temperature, Rainfall, and Severe Local Storm

An interview survey with the villagers was conducted to investigate the perceptions about variations of temperature, rainfall, and severe local storm such as hail and gusty winds associated with developed convective clouds over the last 10 years. The survey was conducted in three periods as follows; March 27–29, 2016 for Period 1, March 8, 2018 for Period 2, and March 7, 2019 for Period 3. The total number of respondents in the surveys was 21. The characteristics of the respondents in the interview survey is shown in Table 4.4. The survey was conducted in the form of a semi-structured interview that delves into the content of perceptions while seeking details.

Table 4.4 Characteristics of the respondents

Sex	Male:19 Female:2
Age class	70s:8 60s:6 50s:4 40s:3
Occupation	Farming:17 Non-farming:4

Table 4.5 shows the perception of the respondents for the trend of temperature. The most common answer is that the annual temperature is rising year by year (43%). The next most common answer is that the temperature of the monsoon is rising (24%). Both perceptions of increasing temperature are consistent with the tendency of annual and monsoon temperature shown in Fig. 4.5.

The number of respondents who recognized the change of temperature is smaller than that of the perception of change of rainfall shown later. Since temperature is not clearly visible unlike rainfall, it may be difficult for many respondents to perceive a change of temperature over a long period (Piya et al. 2012, Vedwan and Rhoades 2001).

Table 4.6 shows the perceptions of the respondents for the trend of rainfall. The most common response is that rainfall during the monsoon is decreasing (43%). The next common response is that rainfall during the pre-monsoon is decreasing (33%). The number of respondents who responded that annual rainfall was decreasing is also relatively high (29%). These perceptions of respondents are consistent with the decreasing trends of rainfall shown in Fig. 4.6. For the post-monsoon, the number of respondents who felt increases and decreases of rainfall were almost the same as 10% and 14%, respectively.

Table 4.5 Perceptions of the trend of temperature

Annual (N = 21)	Increasing:9 Decreasing:2 No change:1 Not stated:9
Pre-monsoon (N = 21)	Increasing:1 Decreasing:0 No change:0 Not stated:20
Monsoon (N = 21)	Increasing:5 Decreasing:0 No change:0 Not stated:16
Post-monsoon (N = 21)	Increasing:2 Decreasing:0 No change:0 Not stated:19
Winter (N = 21)	Increasing:3 Decreasing:2 No change:1 Not stated:15

Source Interview survey

Table 4.6 Perceptions of the trend of rainfall

Annual (N = 21)	Increasing:0 Decreasing:6 No change:1 Not stated:14
Pre-monsoon (N = 21)	Increasing:1 Decreasing:7 No change:0 Not stated:13
Monsoon (N = 21)	Increasing:0 Decreasing:9 No change:0 Not stated:12
Post-monsoon (N = 21)	Increasing:2 Decreasing:3 No change:0 Not stated:16
Winter (N = 21)	Increasing:1 Decreasing:0 No change:0 Not stated:20

Source Interview survey

There were responses that the duration of rainfall had shortened (38%). The relationship between the duration of rainfall and perception for it using rainfall data was not verifiable in this study.

Table 4.7 shows the perceptions of the respondents for the trend of severe local storm and its damage. A notable feature is that many respondents recognized that hail is decreasing year by year (57%), and that the size of the hail had become smaller year by year (38%). For gusty winds, the responses of decreasing and weakening are 29% and 24%, respectively. In Assam, it is known that severe weather such as hail and gusty wind frequently occur in April and May of the pre-monsoon (Mahanta and Yamane 2020) season. Some people mentioned that severe local storms frequently occur during the pre-monsoon season. Murugavel et al. (2012) reported that the instability of the atmosphere during the pre-monsoon is decreasing year by year in northeastern India between 1984 and 2008. Due to the decreasing instability of the atmosphere during the pre-monsoon, the degree of development of convective clouds associated with hail and gusty wind that occur frequently during the pre-monsoon also decreases year by year, which may reduce the frequency of hail and gusts and the size of hail.

Many respondents perceive that the damage caused by hail and gusts is decreasing (19% and 24%, respectively). Some respondents said that this is because the strength of the built structures had increased. Unfortunately, no data on the year-to-year changes in the structure and strength of the building was available, and we were unable to verify the relationship between the decreasing damage by hail and gusty winds and changes in the strength and structure of buildings in Muktapur.

Table 4.7 Perceptions of the trend of severe local storms

Severe local storm (N = 21)	Occurrence	Increasing:0 Decreasing:5 No change:1 Not stated:15
	Damage	Increasing:0 Decreasing:0 No change:1 Not stated:21
Hail (N = 21)	Occurrence	Increasing:1 Decreasing:12 No change:1 Not stated:8
	Size	Larger:1 Smaller:8 No change:1 Not stated:11
	Damage	Increasing:0 Decreasing:4 No change:1 Not stated:16
Gusty wind (N = 21)	Occurrence	Increasing:0 Decreasing:6 No change:1 Not stated:14
	Damages	Increasing:0 Decreasing:5 No change:0 Not stated:16

Source Interview survey

4.6 Conclusion

This study analyzed the relationship between environment and rural livelihoods in Muktapur village, Assam. Changes in the cropping system by introducing modern technologies such as HYV seeds, chemical fertilizers, and agricultural machineries have drastically changed the village's agroecosystem. Indigenous fish and other aquatic species have almost disappeared in paddy fields and surrounding wetlands. Due to the disappearance of certain predatory birds with the loss of aquatic and amphibian species, the monkey population and other animals has drastically increased. Agricultural land use has also undergone considerable changes due to decreasing rainfall, low productivity, labor shortage etc. Some parts of the paddy fields in the village were converted into forestry or fishery ponds, which again caused micro-scale environmental changes in the agricultural fields.

Interactions between environment and local people are constantly occurring in the village. Local people are intimately aware of environmental changes including climate changes and the occurrence of severe local storms. Micro-level environmental changes affect rural livelihoods, especially rice cropping, and the local culture and economy depending on it. In other words, rural livelihoods are deeply embedded into the local environment and ecosystem. The development planners in evolving sustainable strategies and action plans towards improving the condition of agriculture and rural economy in the floodplains of the Brahmaputra valley must understand the value of the traditional cropping system and micro-scale agroecosystem.

References

Asada H (2012) Climate and rice cropping systems in the Brahmaputra Basin: an approach to area studies on Bangladesh and Assam, Rubi Enterprise

Baruah M (2018) Changing status of wetland resources and its impact on livelihood of the wetland communities in Kamrup district Assam, unpublished PhD dissertation, Gauhati University, Assam

Bhagabati AK, Bora AK, Kar BK (2001) Geography of Assam. Rajesh Publications

Bhagabati AK, Deka N (2016) Muktapur Gaon-Paramparagata Krishi aruSanslistaLokasanskriti. Department of Geography, GU and CSEAS, Kyoto University, Japan

Deka N (2012) Agroecosystems in the Brahmaputra Valley, Assam: dynamics and sustainability, PhD dissertation(unpubl.) Gauhati University, India

Deka N, Bhagabati AK (2010) Farming practice in a floodplain village of Assam: continuity and change. Indonesian J Geogr 42(1):13–36

Deka N, Bhagabati AK (2015) Wetlands in a village environment: a case from Brahmaputra floodplain, Assam. Trans Inst Indian Geogr 38(1):35–45

Deka N, Bhagabati AK (2017) Natural resources in a village environment of the Brahmaputra Floodplain. In: Thakur B et al (eds) Regional development-theory and practice, vol 3. Concept publishing Co., Pvt. Ltd, pp 356–380

Deka N, Bhagabati AK, Ando K (2011) Rural land use in the Brahmaputra floodplain environment, Assam: the case of Muktapur village. Contemporary India 1:177–193

Deka RL, Mahanta C, Pathak H, Nath KK, Das S (2013) Trends and fluctuations of rainfall regime in the Brahmaputra and Barak basins of Assam. India, Theoret Appl Climatol 114:61–71

Gopalkrishnan R (2000) Assam: Land and people. Omsons Publications: New Delhi

Government of Assam (2016) Statistical hand book Assam 2016. Directorate of Economics and Statistics, Guwahati

Government of Assam (2017) Economic survey Assam 2017–18. Directorate of Economics and Statistics, Guwahati

Government of India (2017) Agricultural statistics at a Glance 2017. Directorate of Economics and Statistics, Delhi

Kar M (2014) The Brahmaputra: flood and its preventive measures. Eastern Book House

Mahanta R, Yamane Y (2020) Climatology of local severe convective storms in Assam. India. Int J Climatol 40(2):957–978

Murugavel P, Pawar SD, Gopalakrishnan V (2012) Trends of convective available potential energy over the Indian region and its effect on rainfall. Int J Climatol 32(9):1362–1372

Piya, Maharjan KL, Joshi NP (2012) Perceptions and realities of climate change among the Chepang communities in rural Mid-hills of Nepal. J Contemporary India Stud Space Soc Hiroshima Univ 2:35–50

Saikia A (2013) Over-exploitation of forests: a case study from North East India. Springer

Vedwan N, Rhoades RE (2001) Climate change in the Western Himalayas of India: a study of local perception and response. Climate Res 19(2):109–117

Chapter 5
Mercury Rising: Examining LST Links with NDVI, NDWI, NDBI & SMI in the Kamrup Metropolitan District, India Using Geospatial Technologies

Namita Sharma and Gaurav Arote

Abstract Land Surface Temperature (LST) is a basic determinant of terrestrial thermal behavior. It is an important climatic parameter to access the heat of the earth's surface. Soil moisture, water, vegetation cover, and settlement have some strong effects to balance the LST. The heating of the earth's surface is increasing consistently with the passage of time and is a threat to human, animal and plant life. The changes in land cover, especially newly built-up areas create an impact on the variation of LST and soil moisture. Almost all cities of the world are getting warmer today than in the past. The present study analyses the LST of Kamrup Metropolitan district of Assam, along with the Normalized difference water index (NDWI), Normalized difference vegetation index (NDVI), Normalized difference built-up index (NDBI), and soil moisture index (SMI) of the study area for a period of 19 years i.e. 2000 to 2019 using remote sensing (RS) and Geographic information system (GIS) techniques. ENVI 5.1 was used to analyze Landsat satellite imagery, namely Enhanced Thematic Mapper (ETM) & Operational Land Imager (OLI) images. The results of the analysis were imported to ArcGIS 10.2 (www.esri.com) for final layout and map generation. Thermal band (TM) imageries (band 6 & band 10) were additionally used to perform the band math operation to obtain the results. Comparison of all the indices with LST was carried out to ascertain the impact of LST on the environment in the study area.

Keywords LST · NDWI · NDVI · NDBI · SMI · GIS

N. Sharma (✉)
Department of Geography, Gauhati University, Guwahati, India
e-mail: mita199130@gmail.com

G. Arote
Forest Survey of India, MoEF & CC, Dehradun, India

A. Saikia and P. Thapa (eds.), *Environmental Change in South Asia*,
https://doi.org/10.1007/978-3-030-47660-1_5

5.1 Introduction

RS and GIS are emerging tools and technologies that allow us to monitor, investigate, and analyze the changes not only on the earth's surface features but also the atmospheric phenomena using multi-temporal satellite data. GIS plays a vital role in storing, monitoring, mapping, managing, and developing huge amounts of spatial and non–spatial data. Geospatial techniques are widely applied in the observation, detection and analysis of changes on the earth's surface, including climate, environmental degradation, urbanization, industrialization, and its spatio-temporal impact on weather and climate at global and local scales.

Analyzing the anomalies in land surface temperature (LST) is crucial for studies related to global warming and vice-versa. LST is a mixture of vegetation and bare soil temperatures. Vegetation and soil respond frequently to changes in insolation due to cloud cover, aerosol load modifications and diurnal variation of illumination, hence variations in LST can be observed. LST influences the energy between ground and vegetation and is useful for the estimation of air temperature on the earth's surface. LST possesses intrinsic correlations with vegetation and soil system in coastal wetlands (Chi et al. 2020). Land surface reflectance, emissivity, NDVI, soil moisture, and water availability with the amount of rainfall all are interrelated. Degradation of vegetation, deterioration of water bodies and expansion of urban built-up leads to higher LST of an area. Similarly, soil moisture (SM) content affects vegetation cover and surface reflectance. SM is an important component of the Earth's surface water balance and by extension of the energy balance, which regulates the LST and evapotranspiration (Pablos et al. 2016).

The Kamrup Metropolitan district in Assam, India is rich in natural resources, but developmental activities and urbanization have directly impacted the climatic condition and environment of the area. The present study analyzes LST, soil moisture (SM), vegetation index (VI), and built-up index (BUI) to estimate the change over 19 years and assess its impact on the environment. Such studies can be useful for environmental planning and management, conservation of resources, and sustainable development. The health of vegetation cover and water bodies and changes in land cover and built-up area can be assessed using the normalized difference vegetation index (NDVI), BUI, Normalized Difference Water Index (NDWI). SMI influences the geo-biochemical cycle i.e. carbon cycle and water cycle of the earth's surface. Soil moisture is a key parameter that directly or indirectly influences the water cycle (Saha et al. 2018). The NDVI is a simple graphical indicator that is used to analyze remote sensing measurements, targeting the live green vegetation. NDBI can also be used as an indicator of the intensity of development and urban impervious surface. It is feasible to use NDBI to substitute for percent impervious surface area for the study of surface urban heat islands (Macarof and Statescu 2017).

Changes in vegetation cover is mostly due to agriculture, settlement, and industrialization. Such changes impact the water balance of the earth's surface as well as the temperature. Variations in the temperature and water balance affect the properties of soil, hence, variation in soil moisture can be observed. Geoinformatics has the potentiality to evaluate the moisture content, water, healthy vegetation, and surface temperature of an area with minimum technological input. Analysis of multispectral and multi-temporal satellite imageries using RS & GIS enables comparative analyzes at local, national and global levels. Several GIS-based analyzes carried out to evaluate the performance of LST retrieval methods across diverse areas in the mid-latitudes and tropical areas using different land surface emissivity (LSE) models and Landsat missions have been made in recent years. Some studies used the Mono Window Algorithm (MWA), Radiative Transfer Equation (RTE), Single Channel Algorithm (SCA), and Split Window Algorithm (SWA) as methods to retrieve LST of Landsat 5, 7, and 8 (Njoku and Tenenbaum 2022; Abdullah and Barua 2022). In-situ LST data were obtained from the Surface Radiation Budget Network (SURFRAD) stations for accurate evaluation (Sekertekin and Bonafoni 2020). A significant study on soil moisture impact on agricultural drought for Mandri river watershed in Kanker district, Chattisgarh, India was carried out using GIS technique by establishing a relationship between LST and NDVI based on experimental parameterization (Saha et al. 2018). Khandelwal et al. (2018) assessed the variation in LST due to change in elevation of area for Jaipur using MODIS and ASTER data with 8 day's temporal resolution for the year 2006/2007. They used night time LST to avoid direct solar interaction on LST and dependency on the solar zenith angle (Khandelwal et al. 2018), Suresh et al. (2021) correlates the LULC change with LST and conclude that due to the loss of vegetation change from 2001–2019 resulted in temperature increases. Sharath et al. (2021) used lineaments, NDVI, LST, and groundwater drought index (GWDI) to assess drought intensification in the Raichur district. This analysis used MODIS vegetation data and LST to discern drastic changes that occur in water level data in the drought-prone area and GWDI was found to be an effective tool in this context.

Soil moisture mapping for using surface soil samples (0–10 cm) and LST calculation from Landsat7 to identify the cropping pattern in Ismailia governorate, Egypt indicate a high correlation between SMC and SMI (Mohamed et al. 2020). Analysis of LST dynamics with LULC class and its correlation with NDVI & NDBI, including urban–rural gradient analysis using Landsat TM/ETM/OLI images for 1987, 2002, and 2017 reveal a positive relation between LST and NDBI, and a negative relationship between LST and NDVI (Balew and Korme 2020). A change detection method has been used for analyzing NDVI from Landsat data TM/ETM data for the years 2001 & 2006 for Vellore district, Tamilnadu (Gandhi et al. 2015). Assessment of the interrelation between LST, NDVI, and soil moisture content have been carried out to explore the important ecological functions, vulnerable ecosystem, and changes in land surface characteristics in the Yellow River Delta using Landsat 5 data for 1987, 1995, 2005, and Landsat 8 for August and October 2016 and January and April 2017 (Chi, et al. 2020). Macarof and Statescu (2017) made a comparative case study showing NDVI & NDBI as an indicator of surface urban heat island using Landsat data of four seasons for Iasi city in Eastern Romania. Tripathi et al. (2017) estimated the LST of Dindigul city, in Tamil Nadu, India during 1999 to 2016 analyzing the NDVI and BUI using Landsat ETM and OLI data. Xu (2006) modified the NDWI to enhance open water features as it is more suitable for enhancing and extracting water information for a water region with a background dominated by built-up land areas from satellite imageries. The band ratio used for modified normalized difference water index (MNDWI) is Green-MIR/Green + MIR, for Bayi Lake, Fuzhou City. Other studies explored the SM–LST relationship using in-situ and satellite observation over the central part of the river Duero basin in Spain (Pablos et al. 2016).

5.2 Objectives

The purpose of the study is to highlight NDVI, NDWI, BUI, SMI, and LST in the study area for 2000 and 2019 using satellite imagery to analyze the change.

The objectives of the study include:

- To estimate the LST, NDVI, NDWI, NDBI & SMI for Kamrup metro for 2000 & 2019 year
- To establish the correlation between LST with all other indices.

5.3 Study Area

Kamrup metropolitan is a district of Assam, India. The district is bounded by the mighty Brahmaputra River and Darrang district on the north, Meghalaya on the south, Kamrup rural on the west, and Morigaon and Meghalaya on the east. The location of the district is 26.0794 N to 26°11′0″N latitude and 91.6372 to 91° 44′0″E longitudes. The district Kamrup Metro was created by bifurcating the erstwhile Kamrup district on 3rd February 2003. Figure 5.1 shows the Location map of the study area.

Guwahati is a primate city located in Kamrup district. It is an administrative, political, commercial as well as educational hub for Northeast India and a gateway to the seven other neighbouring states of north east India.

The climate is sub-tropical with semi-dry summer & cold in winter. Annual rainfall ranges between 1500 and 2600 mm, average humidity is 76% and temperature varies from a maximum of 37–40° C to 6–7 °C. The district falls under a tropical monsoon rainfall-type climate with high humidity. The topography is a combination of plains and hills.

The area of the Kamrup district is 1527.84 sq. km and the altitude is 55 m from mean sea level (MSL). The district is further divided into six revenue circles, three Community Development blocks, fourteen village councils, and 216 villages. The six revenue circles (RC) are Guwahati RC, Sonapur RC, Chandrapur RC, Dispur RC, North Guwahati RC, and Azara RC. Figure 5.2 Shows the revenue circles of the Kamrup district. A RC is the equivalent of a city ward or municipality found in cities elsewhere across the globe.

The total population of the district is 12, 53,938, out of which 2, 16,927 lives in rural areas and 10, 37,011 is urban population. Brahmaputra is the main river of the

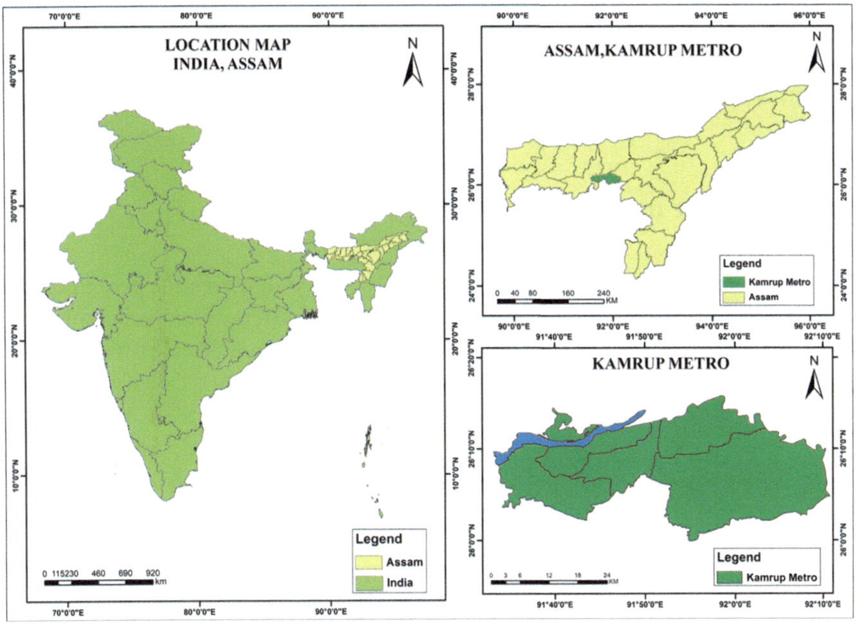

Fig. 5.1 Location map

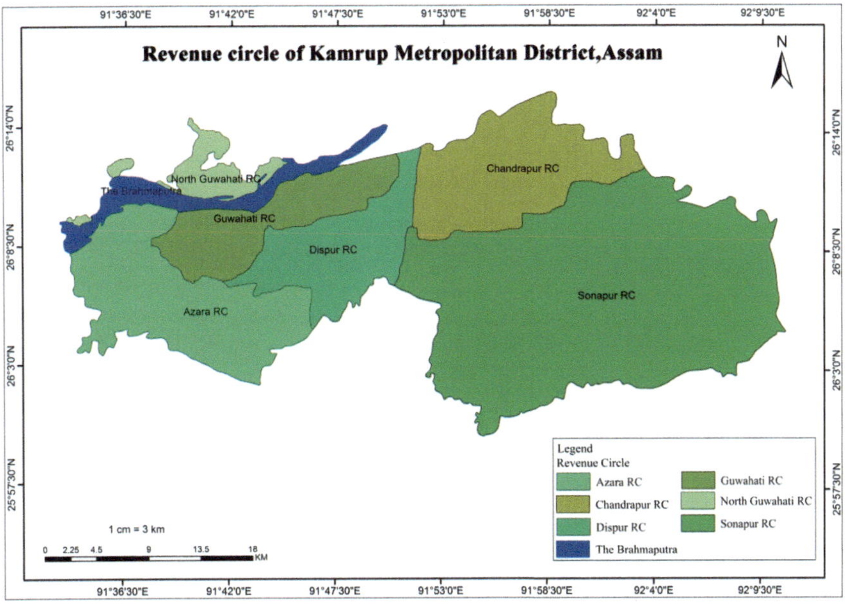

Fig. 5.2 Revenue circle map of Kamrup metropolitan district

district. Bharalu; Digaru and Kolong are some important rivers of the district. The district is connected by several modes of transportation network, which include road, rail, airport, and water ways.

5.4 Database and Methodology

The study area boundary has been created in Q-GIS by digitizing the georeferenced image (censusindia.gov.in) of Kamrup metropolitan districts. Socio-economic data, such as population, literacy rate and percentage of males & females were derived from Kamrup metro's official website (http://kamrupmetro.in/). Details of satellite data used in this research are provided in Table 5.1.

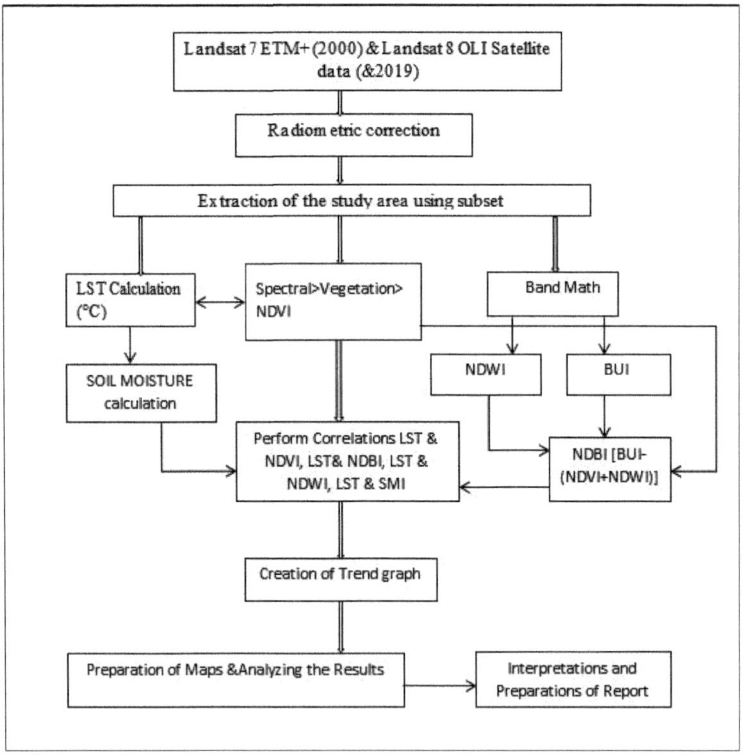

Fig. 5.3 Flowchart showing the methodology adopted in the study

Table 5.1 Details of satellite data used

Satellite	Year	Date	Path/Row	Spatial resolution	Source
Landsat 7 ETM+	2000	21/12/2000 26/11/2000	136/042 137/042	30 m	USGS, Earth Explorer
Landsat 8 OLI	2019	23/01/2019 16/01/2019	137/042 136/042	30 m	USGS, Earth Explorer

5.5 Flowchart of Methodology

To achieve the objectives of the study secondary data were used. Prior to data analysis, radiometric correction (Fig. 5.3) has been performed after downloading the data for further use. The satellite images were mosaicked year-wise and the area of interest for the study was extracted. ENVI 5.1 and Arc GIS 10.2 Software have been used to perform the data processing and analysis. The indices i.e. NDWI, NDBI, BUI were calculated using the band math in ENVI 5.1 platform, using the following formulae:

1. NDVI = (NIR—Red)/(NIR + Red),

   ETM = (Band 4—Band 3)/(Band 4 + Band 3)

   OLI = (Band 5—Band 4)/(Band 5 + Band 4)
2. NDWI = (NIR—SWIR)/(NIR + SWIR)

   ETM = (Band 4—Band 5)/(Band 4 + Band 5)

   OLI = (Band 5—Band 6)/(Band 5 + Band 6)
3. BUI = (SWIR—NIR)/(SWIR + NIR)

   ETM = (Band 5—Band 4)/(Band 5 + Band 4)

   OLI = (Band 6—Band 5)/(Band 6 + Band 5)
4. NDBI = BUI−(NDVI + NDWI)

The application of the individual formula produces values for each indices within −1 to +1. A higher value of NDVI reflects high Near-Infrared (NIR), which represents healthy vegetation cover. NDVI = −1–0 represent water bodies, −0.1–0.1 represent barren rocks, sand, or snow, 0.2–0.5 represent shrubs and grasslands or senescing crops, and 0.6–1.0 represent dense vegetation or tropical rainforest. NDBI and BUI are calculated to analyze the urban expansion. A negative value of NDBI represents water bodies whereas a higher value represents built-up areas. NDWI is used to extract information about the water bodies of a particular area.

5. LST Calculation: Landsat 7 band number '6' and for Landsat 8 band number '10' has been used for LST calculation, using the following method in ENVI 5.1:

5.1. Reflectance from radiometric calibration for all the bands of Landsat 7 and Landsat 8 were calculated in ENVI using the MTL _multispectral file as input.

5.2. NDVI was generated using the reflectance file

Table 5.2 K value for Landsat ETM/OLI

Landsat 7 (ETM)	K1 = 666.09	K2 = 1282.71
Landsat 8 (OLI)	K1 = 774.8853	K2 = 1321.0789

5.3. Proportion of vegetation under each land cover type was calculated using NDVI. Vegetation and bare soil proportions were acquired using the formula:

$P_V = ((b1 - (-NDVI_{min}))/(NDVI_{max}—(-NDVI_{min})))^{2}$, b1 refers to the NDVI file.

5.4. Land Surface Emissivity, which is an important surface parameter dependent on the surface roughness, type and nature of vegetation, surface reflectance etc. was calculated using the formula:

$(\varepsilon) = 0.004 * P_V + 0.986$

5.5. Top of Atmosphere (TOA) was calculated using the MTL_thermal file (TIR) as input. DN was converted to TOA Radiance to produce the output-TIR radiance.

5.6. Brightness temperature (TB) is the radiance of microwave radiation traveling upward from the Earth's surface. TIR_BT was calculated using the formula to generate K value (Table 5.2).

K_2 / (Ln (K_1 / TIR radiance) + 1)) where,

K_1 = Band-specific thermal conversion constant from the metadata

K_2 = Band-specific thermal conversion constant from the metadata

5.6. Conversion of Kelvin to Celsius: b1–273.15, b1 = TIR_BT

5.7. Calculation of LST = (BT/1) + (W *(BT/p)) * Ln (ε))) where,

BT = TIR_BT, For Landsat 7 W is 11.33 (TIR_radiance) Landsat 8 W is 10.900 and p = 14,388 μmk.

P = h*c/s ($1.438*10^{-2}$ mk)

h = Planck's constant ($6.626*10^{-34}$ Js)

s = Boltzmann Constant ($1.38*10^{-23}$ J/k)

c = Velocity of light ($2.998*10^{8}$ m/s)

6. Soil moisture = (LST max-LST)/ (LST max − LST min)

The soil moisture index values range from 0 to 1, where 0 represents extreme dry conditions and 1 represents extreme wet conditions. Soil moisture index can be derived from the LST of the area. After calculating all the results exported to ArcGIS for map preparation.

7. Trend Analysis:

Trend graphs have been represented in the form of scatter plots using x and y coordinates to plot the points. The pattern may reveal a relationship between the values plotted on the grid and can show the trend in data. They are similar to the line graph which helps us to see whether a correlation exists between the variable or not. After creating the fishnet shape file under data management tool on ArcGIS 10.1, the values of each and every index were added to the fishnet point file using multi values to point under spatial analysis tool to plot the data on the scatter plot. Trend graph for all the indices obtained from the scatter plot to identify the correlations that exist among the variables.

5.6 Results and Discussion

LST, NDVI, NDWI, SMI & NDBI are useful indices to extract information related to surface temperature, vegetation cover, water availability, soil moisture, and area under built-up and barren land respectively. These indices are used to collect information regarding the changes that had accrued over the last two decades. The thermal infrared band was converted to spectral radiance i.e. TOA and BT using ENVI 5.1. The LST maps shown in Figs. 5.4 and 5.5 for the years 2000 to 2019 respectively indicate higher temperatures in the built-up and other impervious areas, while lower temperatures are recorded in the forested or vegetative areas. The lowest radiant temperature derived from the analysis of LST 2000 is 14.9 °C in the forest area, and the highest radiant temperature is 28 °C in the built-up area (Fig. 5.4). However, in the LST map of 2019

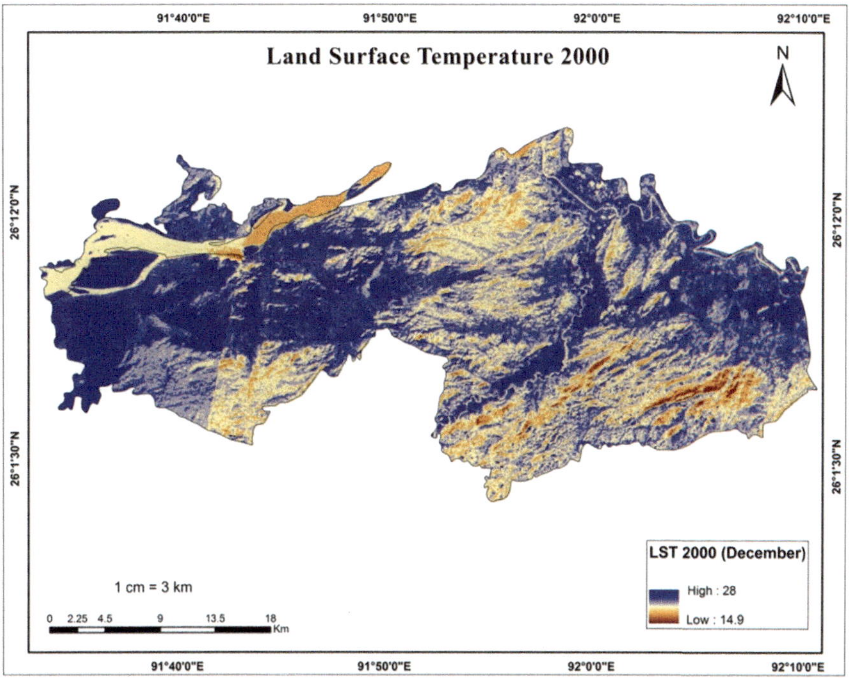

Fig. 5.4 LST of 2000

the lowest and highest radiant temperature range to be 14.6 °C to 29 °C respectively. This study further attempts to compare water, forest, agriculture, vegetation, built-up area, barren land, and soil moisture with the surface temperature of both years. Special emphasis is given to compare the temperature distribution with different aspects. The implication of urban development by replacing natural vegetation (forest) with built-up surfaces such as concrete, stone, metal, and asphalt leads to an increase in the surface temperature.

The highest value of NDVI for 2000 is 0.59 (Fig. 5.6), and for 2019 it is 0.48 (Fig. 5.7). These values indicate a negative change in terms of vegetation cover. NDVI values varied from 0.1 to 0.59. Less vegetated areas, including built-up areas and water bodies, have the lowest NDVI values. Since the reflection from the soil is high and it produces low values in the NIR band as well as the same for the built-up area and produces high values in the red band. As water absorbs more energy, the bands get darker values resulting in less reflectance. The highest values for both the years 2000 and 2019 i.e. 0.48 and 0.59 respectively indicate the density of green healthy vegetation, which are mainly distributed in less populated areas. However, with an increase in built-up area between 2000 to 2019, the density of green vegetation has decreased, in a rather predictable manner.

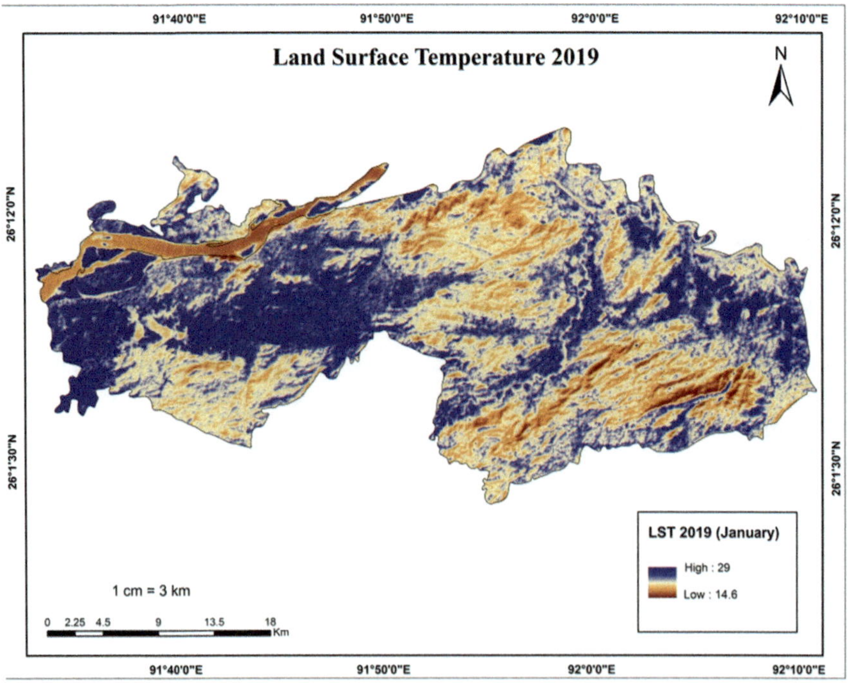

Fig. 5.5 LST of 2022

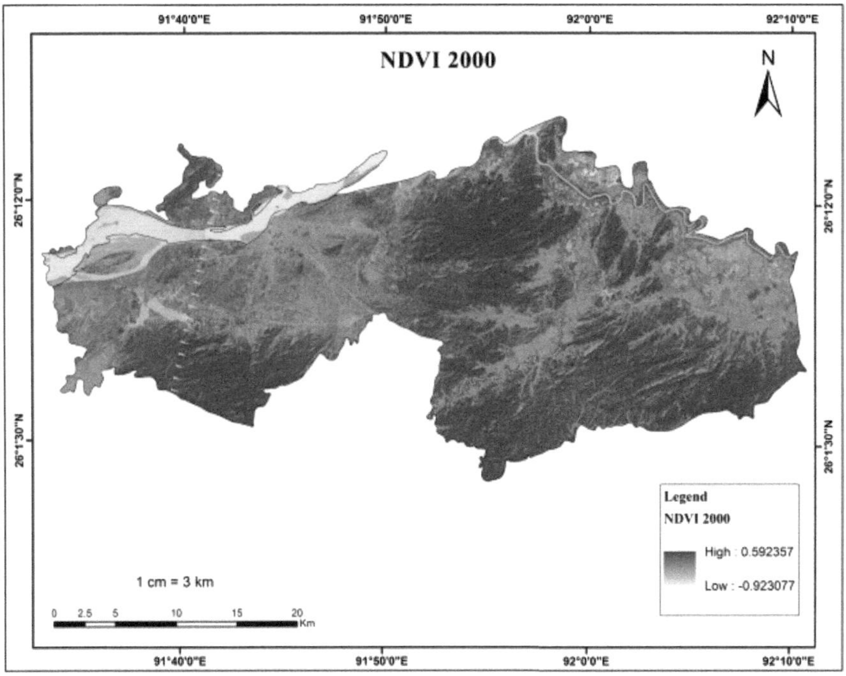

Fig. 5.6 NDVI of 2000

The value of water index also depends on the summer and winter seasons. Results of NDWI generated values ranging from 0.66 to −0.91 for the year 2000 (Fig. 5.8) and it varies from 0.31 to −0.292 for the year 2019. The variation in the reflectance value of NDWI, which is related to the rainy and dry seasons can be attributed to the influence of many seasonal wetlands in the study area. NDWI values vary from month to month. The NDWI value has been taken for December (2000) and January (2019) depending on the availability of cloud-free data, and less rainfall compared to the month of May and June, which is the peak monsoon season. NDWI separates the water bodies from the other forest area and built-up areas. A value that is greater than zero indicates the presence of water bodies and water-logged areas. NDWI values are considerably lower in 2019 compared to those existing in 2000. The data for both 2000 and 2019 were during the winter season, and the results indicate a decreasing trend in the water bodies (Fig. 5.9).

Remotely sensed imagery is ideally used to monitor and detect land cover changes that occur frequently in urban and peri-urban areas as a consequence of growing urbanization. In this study, the Normalized Difference Built-up Index (NDBI) has been calculated using band math to automate the process of mapping built-up areas. It takes advantage of the unique spectral response of built-up areas and other land covers. NDBI index captures urban areas or built areas which usually record higher reflection in the Shortwave Infrared (SWIR) area compared to Near-Infrared (NIR)

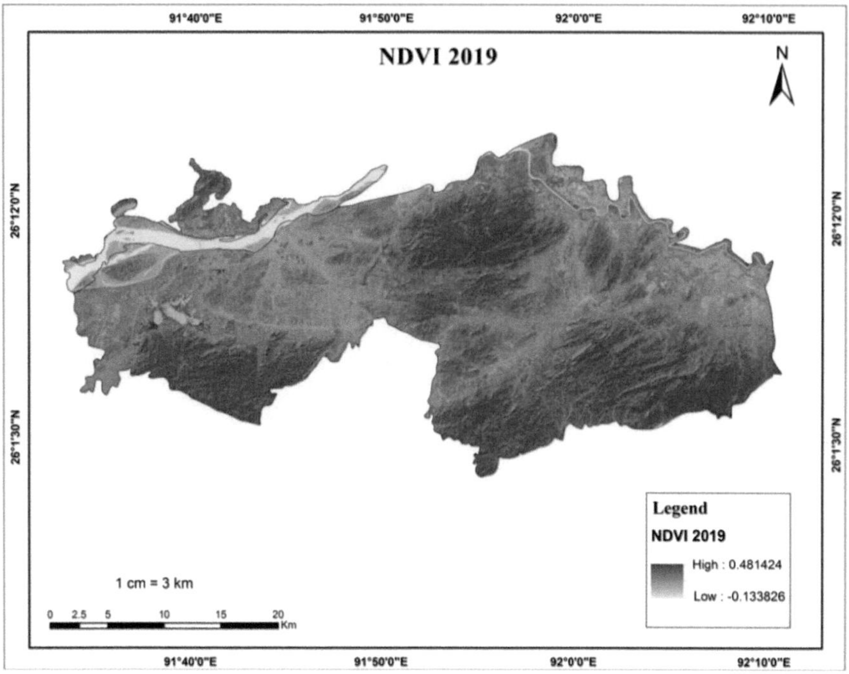

Fig. 5.7 NDVI of 2019

spectrum. These applications are very useful for monitoring and planning land use/land cover (LULC). Figure 5.10 NDBI and Fig. 5.11 show the built-up areas in red color which represents the highest value. The NDBI values vary between −0.7 to 0.42 in 2000 and −1 to 0.49 in 2019. Major built-up areas are found in the Western part of the study area as it is the city region of the area. The lowest value is recorded in the hilly and forested areas. Built-up areas exhibit higher heat conductivity than bare soil, resulting in higher emissivity and albedo values of thermal infrared radiation in built-up areas relative to bare soil areas. The region with higher spectral values are mostly occupied by settlement and bare soil, while lower spectral values are indicative of areas dominated by vegetation and water bodies.

The results of the soil moisture index map of 2000, which has been grouped into five classes are within the range of 0 to 1. Most of the study area, as shown in Fig. 5.12 (shaded in light brown to dark brown color), has a value close to zero, which indicates that these areas are highly affected by water deficit. The values closer to 1 (dark brown color) are the areas that have higher moisture content compared to the rest of the land use categories. The results of 2000 reveal higher soil moisture values indicating that these areas are covered with good vegetation and water availability during that year. The results of 2019 (Fig. 5.13) reveal that more than 60–70% of the area has values close to zero, which means that these areas are currently moisture

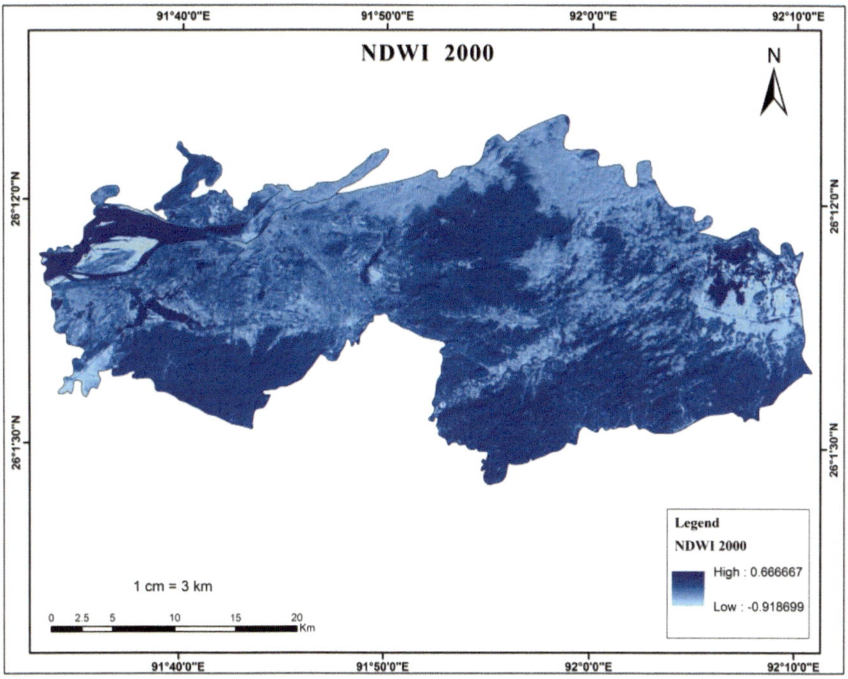

Fig. 5.8 NDWI of 2000

deficit. SMI index ranges from, 1 (representing higher presence of water or moisture content), and zero (indicating lower moisture content or dry areas).

5.7 Trend Analysis

Trends analysis (Figs. 5.14, 5.15, 5.16 and 5.17) shows the relationship between the LST and various other indices. Scatter plot graph is used to analyze the correlation and trend among the data. Comparison of various indices with LST clearly shows the variations as the pixel values are more irregular in 2000 than in 2019. LST & NDVI, NDWI, and SMI are negatively related, while NDBI & LST are positively related. LST & NDVI has an indirect relationship as LST values increase with decreasing NDVI values. Similar relationship can be observed between LST & NDWI and LST & SMI. A direct relationship is established between NDBI & LST as the value of LST increases with the increase in the values of NDBI. Trend analysis for the year 2000 depicts a near to uniform pattern or systematic pattern than that of 2019, which displays scattered or random patterns indicating the expansion of urban areas

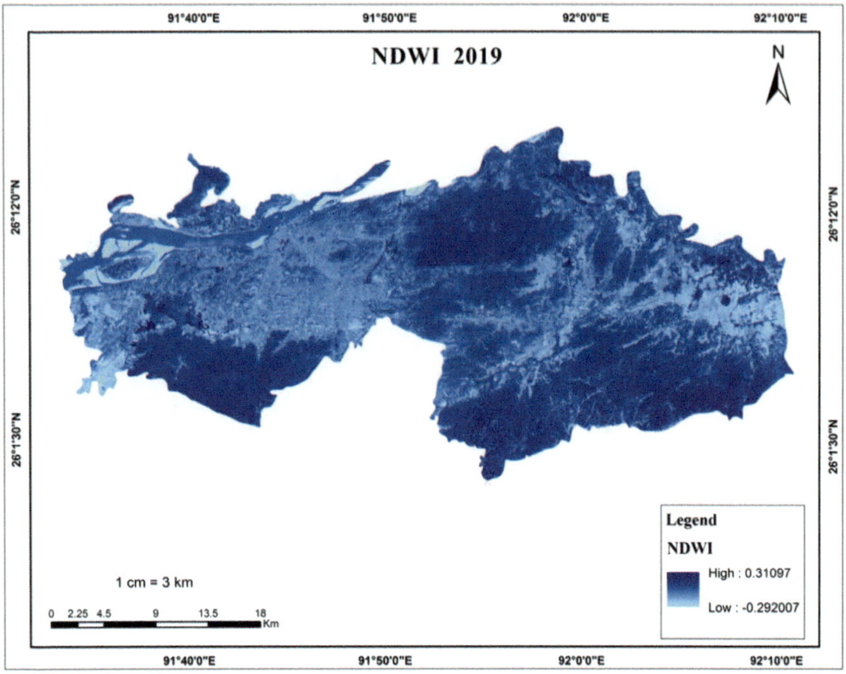

Fig. 5.9 NDWI of 2019

and changes in the use of land. LST has a strong negative correlation with NDVI & NDWI, hence areas with the least water content and vegetation cover area experience a higher value for LST. Soil moisture depends on LULC, As such, an increase in vegetation cover also increases soil moisture content. However, increases in built-up areas due to anthropogenic drivers tend to induce soil degradation leading to moisture deficit. As temperature increases, the water stored in the soil gets evaporated and the insufficient moisture present in the soil in turn adversely affects plant growth and agricultural yield.

5.8 Conclusions

Analysis of the change in LST in relation with NDVI, NDWI, SMI, & NDBI of Kamrup metropolitan district for 2000 and 2019 reveals an increase of 1 °C in LST over a period of 19 years. This can be mainly attributed to the expansion of built-up areas and reduction in water bodies and vegetation cover, which ultimately leads to variations in LST. Thus, the degradation of vegetation and water bodies and expansion of the urban areas are the major causes of the change in surface temperature. LST is also a major factor responsible for the urban heat island effect. Analysis of

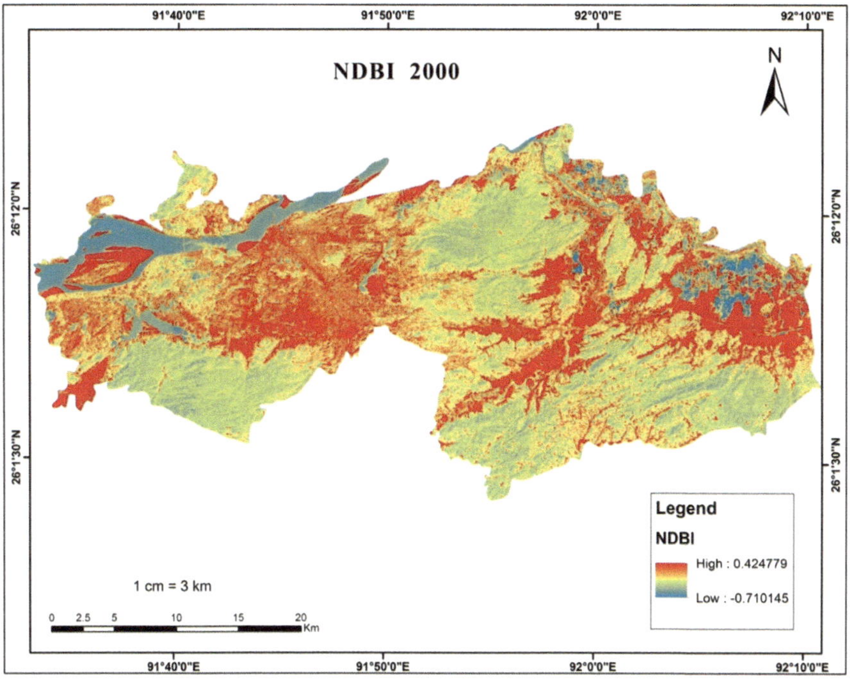

Fig. 5.10 NDBI of 2000

LST, NDVI, NDWI, SMI and NDBI becomes necessary as the urban and sub urban area are continuously expanding and cause various negative impacts on the environment. The results of this study was verified by conducting a trend analysis so as to ascertain the correlation of LST with all the individual indices. The analysis further underscores the fact that an increase in built-up areas leads to a decrease in vegetation cover and water bodies, which ultimately results in an increase in LST. Urban and suburban areas recorded high surface temperatures, while the non-built-up and vegetative areas exhibited higher NDVI values. Such trends along with soil moisture deficit adversely affect the climate and environment on a long–term basis. The positive change in NDBI is indicative of the expansion and development of Gauhati city. Unfortunately, urban sprawl (Pawe and Saikia 2020), and increases in LST (Chetia et al. 2020) that Guwahati has experienced in recent years and negative changes in NDVI, NDWI, and SMI discerned by this analysis, raises certain environmental concerns, which can deleteriously affect the city's sustainability in the long run. The urban situation in Guwahati is not dissimilar to other smaller urban centers such as Imphal (Neog 2021), Dimapur (Neog 2022a, b) and Agartala (Neog 2022a, b) in Manipur, Nagaland and Tripura respectively. Skewed weather conditions, variations in the summer and winter seasons and extreme weather events that are being increasingly witnessed the world over as a result of climate change could aggravate Kamrup metropolitan center's conditions. Introducing green spaces within the city through

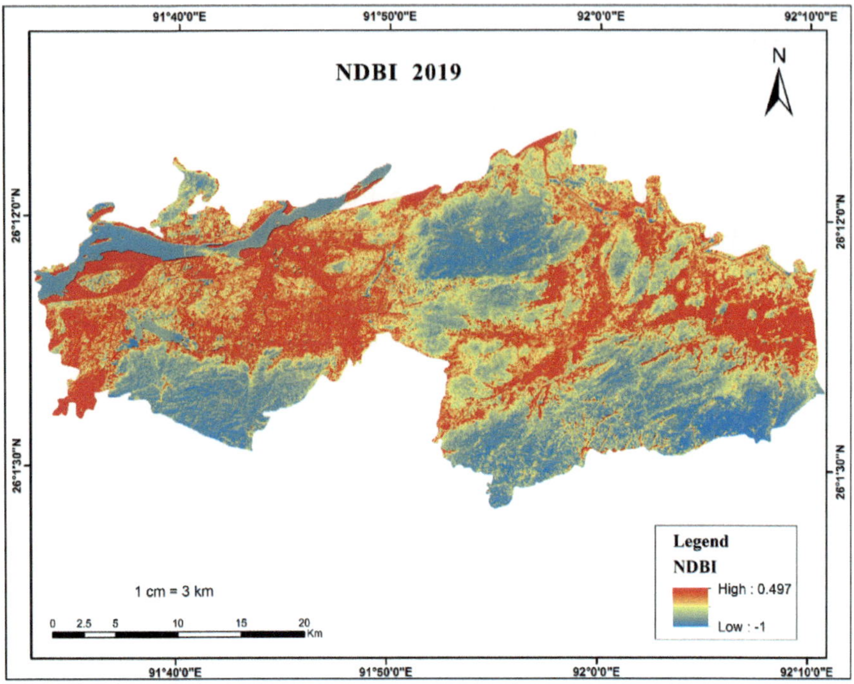

Fig. 5.11 NDBI of 2019

afforestation along the roadside, community parks, and protection of eco-sensitive can be some of the effective measures to reverse and control the unfavourable patterns and trends resulting from the analysis of LST and all the indices in the present study. Since a recent analysis shows a positive relationship between LST and built-up areas in Guwahati (Chetia et al. 2020), maintaining and improving the quantum of green areas is of prime importance (Pawe and Saikia 2020). Moreover, proper planning and management strategies must be adopted to check unplanned growth of cities, minimize environmental degradation, reduce climate change impacts, and maximize ecosystem services. In seeking to monitor environmental parameters in the urban landscape of Guwahati and the Kamrup metropolitan district taking recourse to measures such as crowdsourcing data and analysis (See et al. 2016) along with other smart city initiatives need also be considered. In these, the advantages of using geospatial datasets and methods are undeniable.

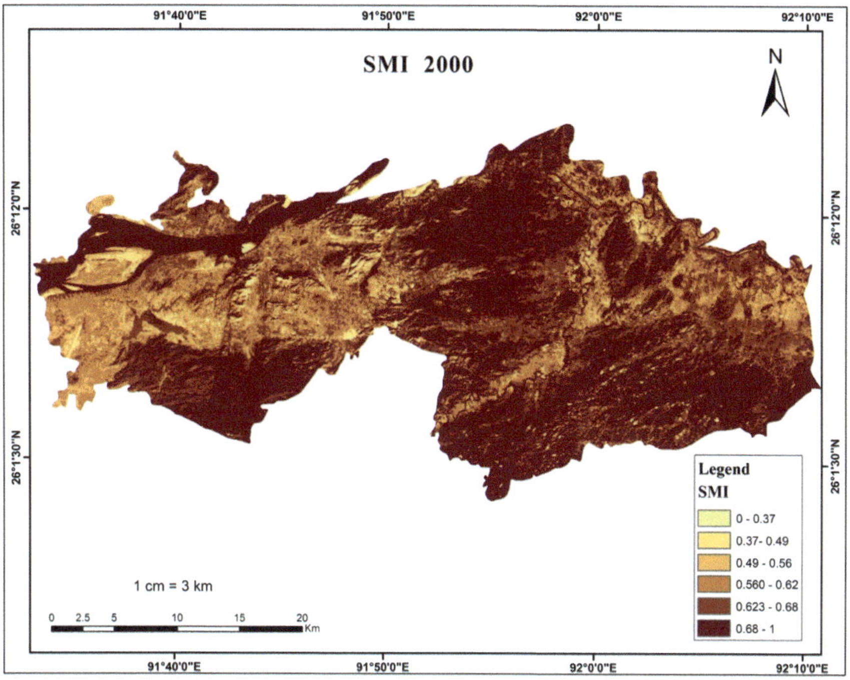

Fig. 5.12 SMI of 2000

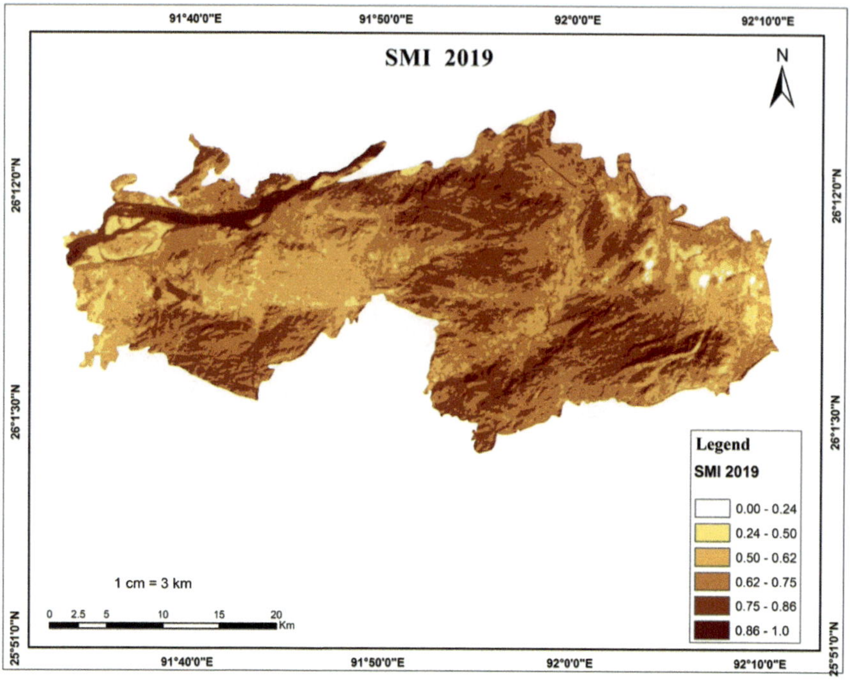

Fig. 5.13 SMI of 2019

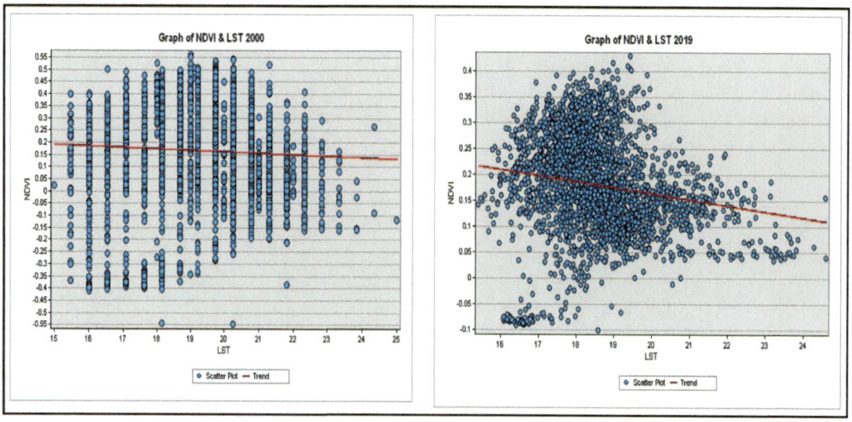

Fig. 5.14 Trends of LST & NDVI of 2000, 2019

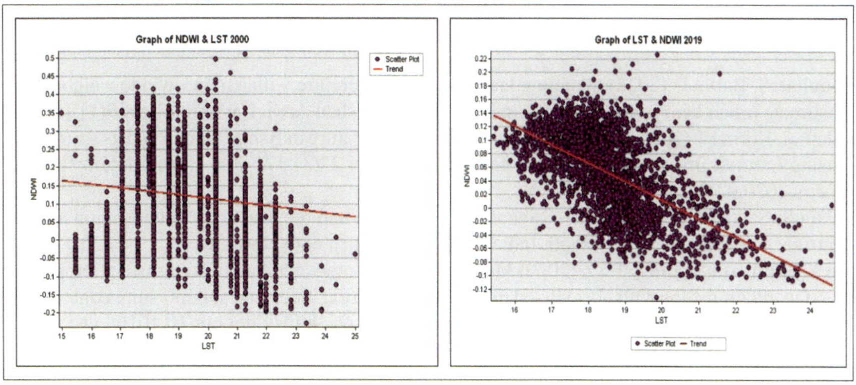

Fig. 5.15 Trends of LST & NDWI of 2000, 2019

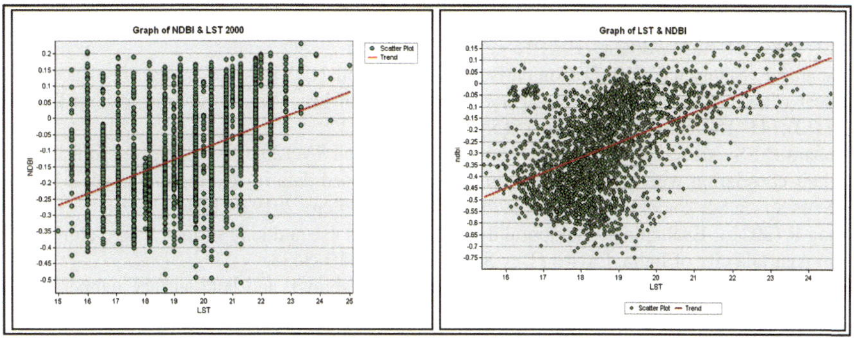

Fig. 5.16 Trends of LST & NDBI of 2000, 2019

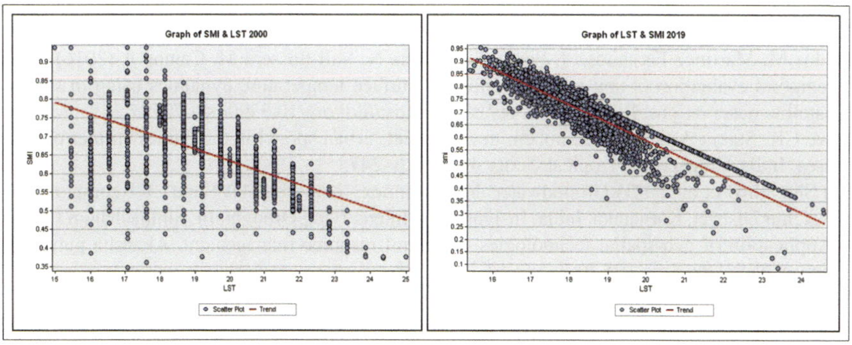

Fig. 5.17 Trends of LST & SMI of 2000, 2019

References

Abdullah S, Barua D (2022) modeling land surface temperature with a mono-window algorithm to estimate urban heat Island intensity in an expanding Urban Area. Environ Proces 9(1):14

Balew A, Korme T (2020) Monitoring land surface temperature in Bahir Dar city and its surrounding using Landsat images. Egypt J Remote Sens Space Sci 23(3):371–386. https://doi.org/10.1016/j.ejrs.2020.02.001

Chetia S, Saikia A, Basumatary M, Sahariah D (2020) When the heat is on: urbanization and land surface temperature in Guwahati. India. Acta Geophysica 68(3):891–901

Chi Y, Sun J, Sun Y, Liu S, Fu Z (2020) Multi-temporal characterization of land surface temperature and its relationships with normalized difference vegetation index and soil moisture content in the Yellow River Delta. China. Global Ecol Conser 23:e01092. https://doi.org/10.1016/j.gecco.2020.e01092

Gandhi GM, Parthiban S, Thummalu N, Christy A (2015) Ndvi: vegetation change detection using remote sensing and GIS—a case study of Vellore District. Procedia Comput Sci 57:1199–1210. https://doi.org/10.1016/j.procs.2015.07.415

Khandelwal S, Goyal R, Kaul N, Mathew A (2018) Assessment of land surface temperature variation due to change in elevation of area surrounding Jaipur, India. Egypt J Remote Sens Space Sci 21(1):87–94. https://doi.org/10.1016/j.ejrs.2017.01.005

Macarof P, Statescu F (2017) Comparison of NDBI and NDVI as indicators of surface urban heat island effect in landsat 8 imagery: a case study of Iasi. Present Environ Sustain Develop 11(2):141–150. https://doi.org/10.1515/pesd-2017-0032

Mohamed ES, Ali A, El-Shirbeny M, Abutaleb K, Shaddad SM (2020) Mapping soil moisture and their correlation with crop pattern using remotely sensed data in arid region. Egypt J Remote Sens Space Sci 23(3):347–353. https://doi.org/10.1016/j.ejrs.2019.04.003

Neog R (2022a) Evaluation of temporal dynamics of land use and land surface temperature (LST) in Agartala city of India. Environ Dev Sustain 24(3):3419–3438

Neog R (2022b) Monitoring land use dynamics, urban sprawl, and land surface temperature in Dimapur urban area, Nagaland, India. Int J Environ Sci Technol

Neog R (2021) Analyzing dynamic behavior of land use and land surface temperature in the city of Imphal, India. Acta Geophysica 69(6):2275–2290

Njoku EA, Tenenbaum DE (2022) Quantitative assessment of the relationship between land use/land cover (LULC), topographic elevation and land surface temperature (LST) in Ilorin, Nigeria. Remote Sens Appl Soc Environ 27:100780

Pablos M, Martinez Fernandez J, Piles M, Sánchez N, Vall-llossera M, Camps A (2016) Multi-temporal evaluation of soil moisture and land surface temperature dynamics using in situ and satellite observations. Remote Sens 8(7):587. https://doi.org/10.3390/rs8070587

Pawe CK, Saikia A (2020) Decumbent development: urban sprawl in the Guwahati metropolitan area. India. Singapore J Tropical Geogr 41(2):226–247

Raj Sharath B et al (2021) Groundwater drought index, temporal NDVI and LST comparison of Raichur District, Karnataka, India: Insight to the Intensifying Drought. In: Simhachalam DA (ed) Application of geo-spatial technologies in natural resource management. Akansha publishing house, New Delhi, pp 105–117

Saha A, Patil M, Goyal VC, Rathore DS (2018) Assessment and impact of soil moisture index in agricultural drought estimation using remote sensing and GIS techniques. Proceedings 7(1):2. https://doi.org/10.3390/ecws-3-05802

See L, Fritz S, Perger C, Changkakati T, Obersteiner M et al. (2016) Mapping human impact using crowdsourcing. In: Carver SJ, Fritz S (eds) Mapping wilderness: concepts, techniques and applications. Dordrecht, Springer, pp 89–101

Sekertekin A, Bonafoni S (2020) Land surface temperature retrieval from landsat 5, 7, and 8 over Rural areas: assessment of different retrieval algorithms and emissivity models and toolbox implementation. Remote Sens 12(2):294. https://doi.org/10.3390/rs12020294

Suresh DS et al (2021) LULC changes and LST dynamics in Davangere city, Karnataka India: inferences from landsat data. In: Simhachalam A (ed) Application of geo-spatial technologies in natural resource management. Akansha publishing house, New Delhi, pp 235–245

Tripathi R, Rahman MP, Abdul S (2017) Estimation of land surface temperature using geoinformatics technique, a case study of Dindigul city, Tamilnadu, India. In: Muthukumer DM (ed) Geospatial technology for rural development. Shanlax publications, Madurai, pp 133–138

Xu H (2006) Modification of normalized difference water index (NDWI) to enhance open water features in remotely sensed imagery. Int J Remote Sens 27(14):3025–3033. https://doi.org/10.1080/01431160600589179

Chapter 6
Long-Term Shoreline Change Analysis and Impact of Hard Coastal Structures on Shorelines—A Case Study of the Nagapattinam-Ramanathapuram Shoreline

C. Prakasam, R. Aravinth, and S. Sanjeevi Prasad

Abstract Shorelines are dynamic and complex in nature and shoreline changes are induced by numerous factors such as littoral currents, waves, riverine activities and sometimes tides as well. Construction of hard coastal structures disrupts natural deposition tendencies and erosion occurring along the coast. The way they are affected depends upon the type of structure and longshore transport. This Chapter examines a shoreline stretch of approximately 376 kms from Kodikkarai in (Nagapattinam) to Mandabam (Ramanathapuram) covering the districts of Thanjavur, Thiruvarur and Pudhukottai in Tamil Nadu, India. The shoreline change analysis entailed the use of satellite imageries used are derived from LANDSAT (MSS, TM, ETM, ETM+ and OLI) during the years 1988–2015. To derive the shoreline data from the satellite images, band 5 in Landsat (MSS) and band 6 (ETM–OLI) were used as they allow us to distinguish between land and water boundaries clearly. The shorelines were demarcated and digitized using various sources including LANDSAT, LISS and Google Earth Imageries. The datasets were saved based on the standard format of (DD/MM/YYYY) for analysis. The erosion and accretion rates have been calculated using the DSAS plugin. The resultant maps were used to demarcate depositional features that were predominantly observed over the period 1988 to 2015. The highest amount of deposition and erosion were along Tirutturappundi and Mandabam respectively. For, the assessment of the impact of hard structures along the coast requires high resolution satellite imageries, hence Google Earth imageries were used to interpret the data. Most of the sites show both depositional and erosional features. Site

C. Prakasam (✉)
Department of Geography School of Earth Sciences, Assam University, Diphu Campus (A Central University), Diphu, Karbi Anglong, Assam 782462, India
e-mail: cprakasam@gmail.com; c.prakasam@aus.ac.in

R. Aravinth
Institute of Environment Education and Research, Bharathi Vidyapeeth University, Pune 411043, India

S. Sanjeevi Prasad
Department of Geography, University of Madras, Chennai 600005, Tamil Nadu, India
e-mail: sanjeevi@unom.ac.in

A. Saikia and P. Thapa (eds.), *Environmental Change in South Asia*,
https://doi.org/10.1007/978-3-030-47660-1_6

1 had only erosional features, while site 14 had depositional features all along the coastline.

Keywords Net shore movement · End point rate · Shoreline change · DSAS · Remote sensing

6.1 Introduction

Shoreline change is an ongoing process over millions of years. A beach is an environment that is present between land and sea and is a fragile ecosystem and constantly changing due to various conditions (Kaliraj et al. 2017). Landforms are modified by both natural and man-made processes (Kankara et al. 2015; Lakshmanna and Prasad 2018). Changes induced by human intervention tend to alter these natural processes. Construction of harbours, jetties and sea walls have influenced erosion and accretion patterns (Fig. 6.1). Shoreline change analysis is a step towards assessing the vulnerability of coasts and the dominant factors that play a role in it (Dwarakish and Nithyapriya 2016; Kumar et al. 2010). Estimating the causative factors helps in taking precautionary measures while designing coastal structures in future. Coastal structures are constructed to trap longshore sediment to create protective beaches. These structures are found to cause deposition along the up-drift side and erosion along the down-drift side (Mahendra et al. 2011; Maiti and Bhattacharya 2009).

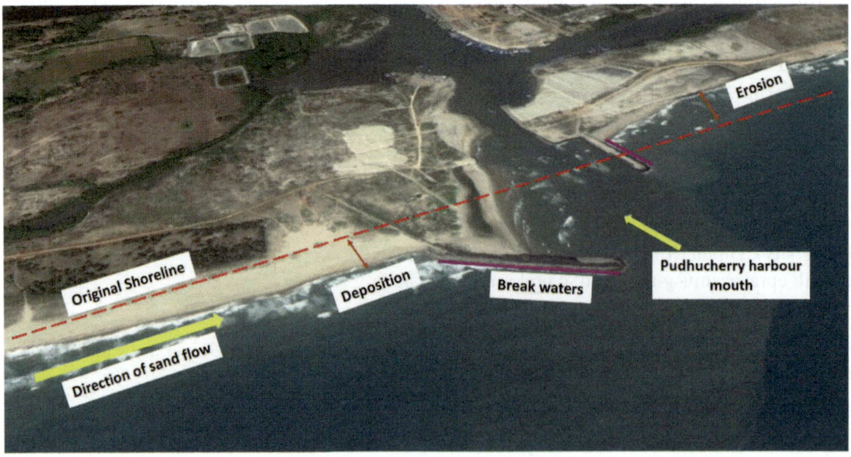

Fig. 6.1 Effect of hard coastal structures on a shoreline

Numerous scientists and researchers have studied shoreline changes along the Tamilnadu coastline (Natesan et al. 2015; Saravanan et al. 2019; Vivek et al. 2019) and the impacts of coastal structures like harbours and ports along a coastline.

In the present work, ERDAS Imagine, ArcGIS and DSAS tools have been used to calculate the changes along the coast. Layer stack and pan merge techniques were used to interpret satellite imageries. End point rate (EPR), Net shoreline movement (NSM) and Weighted linear regression (WLR) models have been used to identify erosion, accretion rates and map vulnerable areas.

Objectives

The study focuses on assessing the changes along the coastline of Nagapattinam to Ramanathapuram district and also to assess changes in the shoreline induced by the construction of hard structures along this stretch of coastline. We considered the following:

1. Demarcation of shoreline for various years through LANDSAT and Google earth imageries.
2. Estimating erosion and accretion output for EPR, NSM and WLR.
3. Mapping of highest deposition and erosion through EPR methods for long term analysis and changes induced by hard coastal structures.

6.2 Study Area

The study area extends from Kodikarai in Nagapattinam to Rameshwaram in Ramanathapuram district (Fig. 6.2) of Tamil Nadu, India. The entire study covers about 560 kms of coastline comprising the five coastal districts of Nagapattinam, Thiruvarur, Pudhukottai, Thanjavur and Ramanathapuram. This stretch of coastline is characterized by the presence of coastal structures like jetties and dockyards.

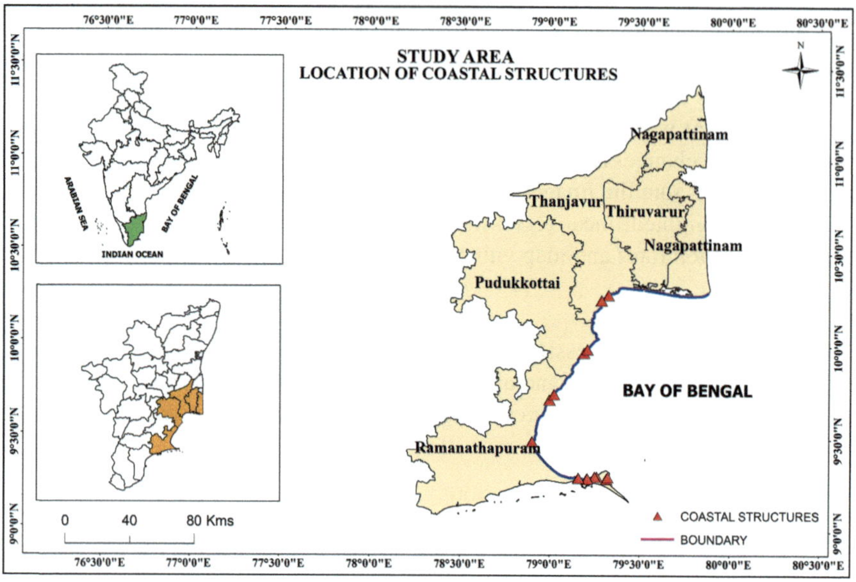

Fig. 6.2 Study area

6.3 Materials and Methods

6.3.1 Data Used

Multi-temporal satellite data from LANDSAT (MSS, TM, ETM and OLI) were used. Data for the years 1988, 1992, 1995, 2000, 2005, 2010, and 2016 were used to delineate the shoreline (Table 6.1). Mapping of coastal structures and shorelines along them were achieved using multi-temporal satellite data from Google Earth. The total number of years taken into account for the study area was determined by the availability of Google Earth data.

Table 6.1 Data used for analysis

Sl. no	Data	Year	Source E	Purpose
1	Landsat TM	1992, 1995, 2005,2011	USGS Earth Explorer	Shoreline extraction
2	Landsat 7 ETM	2000	USGS Earth Explorer	Shoreline extraction
3	Landsat 8 OLI	2015	USGS Earth Explorer	Shoreline extraction
4	Google Earth imageries	2000–2016	Google Earth	Shoreline extraction and coastal structures

6.3.2 Methodology

Band 5 in Landsat 5 & 7 and Band 6 in Landsat 8 use short wave infra-red that clearly facilitates differentiating between land and water. These bands were used to demarcate the shoreline in the satellite imageries. The digitized shorelines were then assigned respective dates [mm/dd/year]. Thereafter shorelines were exported into a personal database in a GIS environment using ArcGIS (www.esri.com). The entire coastline was divided into 100 m transec

ts to calculate shoreline change. The Digital Shoreline Analysis System (DSAS) which is an ArcGIS extension created by USGS was used to analyse the shoreline change rate. The end point rate, net shoreline movement and weighted linear regression were calculated and the results were then exported as a table (Fig. 6.3). These tables were then used to create maps to identify erosional and depositional areas throughout the study area (Sowmya et al. 2019; Sreenivasulu et al. 2016; Vivek et al. 2019).

Analysing the impacts of hard coastal structures along the shoreline requires high resolution satellite imagery. Hence, multi–temporal data from Google Earth were used to identify various coastal structures and delineate shorelines along these structures. A total of 18 structures classified into 14 sites were considered in this study. The year of the study chosen was different for every structure as it was based on the availability of data for that particular structure. The coastal stretch taken along the structures varied from 300 to 1000 m in length. Satellite imageries from Google Earth were found to have geometric errors which were corrected after digitizing all the shorelines. Ground control points (GCP) were collected for all the sites and shorelines were corrected with respect to the acquired GCP points. The digitized shorelines were then assigned respective dates [mm/dd/year] and the shorelines were then exported into a personal geodatabase in ArcGIS environment. The shorelines were then divided into 1 m transects to calculate shoreline change and outputs from the DSAS software were calculated and the results exported as a table (Fig. 6.3). These tables were then used to create maps to identify erosional and depositional areas throughout the study area (Saleem and Awange 2019; Sankari et al. 2015; Saxena et al. 2013).

6.4 Result and Discussion

6.4.1 Assessment of Long—Term Change in Shorelines

Long-term shoreline change was estimated from Kodikarai in Nagapattinam to Mandabam in Rameshwaram district through visual image interpretation and shoreline delineation for the years (1988 to 2015) of post–monsoon season. The output was differentiated into three types namely erosional and depositional features and no

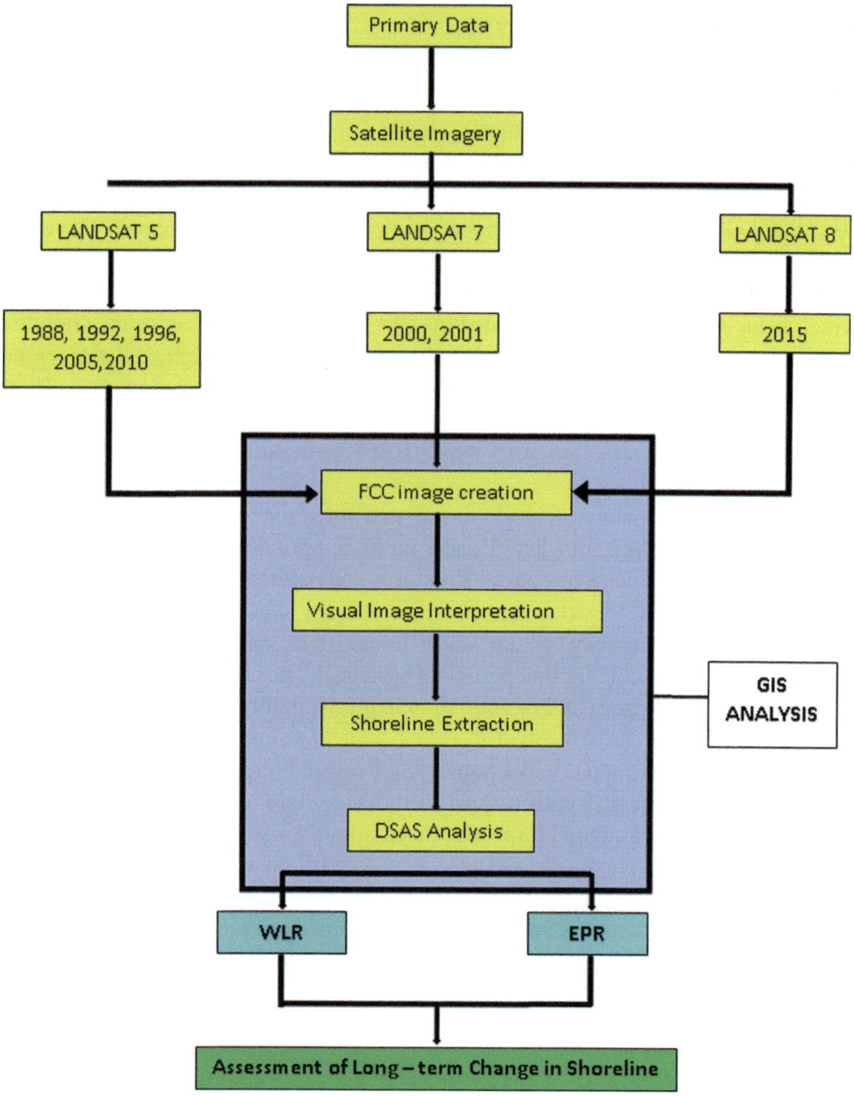

Fig. 6.3 Overall research methodology

change. The entire shoreline was categorized into two sections as they spread into two different Landsat scenes.

Section 6.1 extends from Kodikkarai in Nagapattinam to Pathanendal in Ramanathapuram district (Fig. 6.4). The total shoreline extent is 336 kms covering Nagapattinam to Pathanendal of Rameshwaram. A total of 6 years were taken for this study (1988, 1996, 2000, 2005, 2010 and 2015) over a span of 27 years. Shoreline change was calculated between each year and overall change was calculated during 1988 to 2015. The results were generated and outputs were presented as maps. **Section** 6.2 is located in Ramanathapuram taluk (Fig. 6.5). The total shoreline extent is 40 kms covering the entire Ramanathapuram taluk. A total of 6 years were taken for this study (1988, 1996, 2001, 2005, 2010 and 2015) across a span of 27 years.

Along Sect. 6.1, during 1988 to 1995 alternating erosion and depositional shorelines were formed. The highest deposition and erosion were identified along the coast of Tirutturappundi and Peravurani. Most depositional features were found along the coast of Pattukotai, Mannargudi and Tiruttarappundi. Between 1995 and 2000 alternating erosion and depositional shorelines were formed, while peak deposition and erosion was identified along the coast of Mannargudi and Arantangi. Most of the erosional features were found along the coast of Tiruttarappundi, Mannargudi,

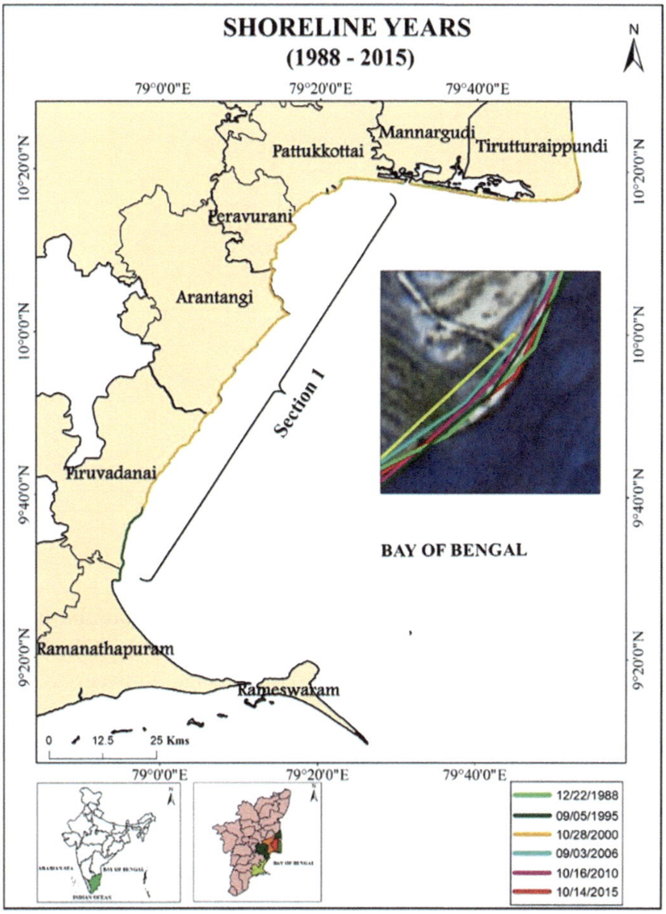

Fig. 6.4 Section 6.1 of study area

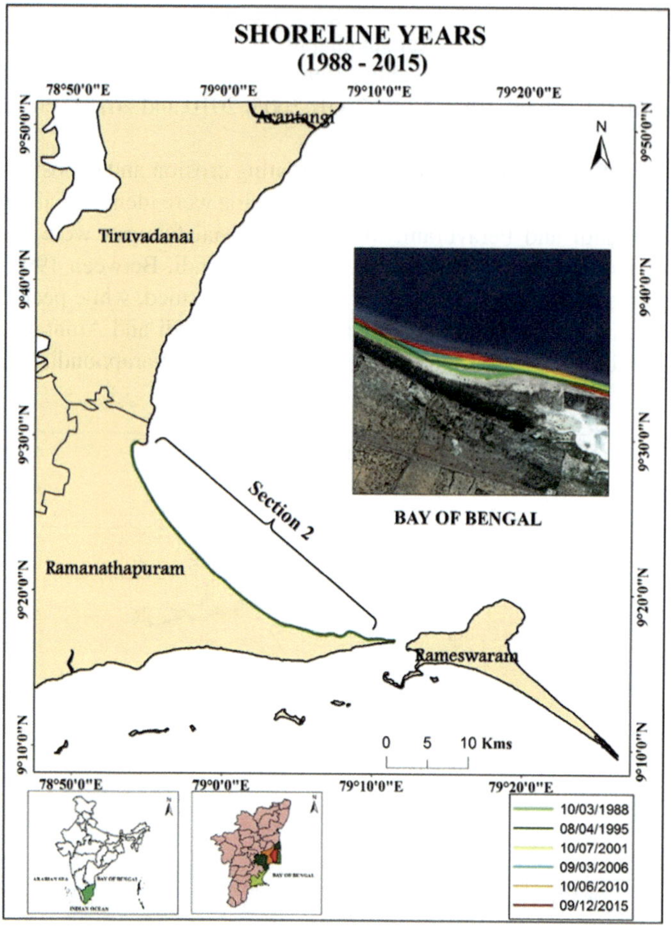

Fig. 6.5 Section 6.2 of study area

Pattukotai, Peruvanai and Arantangi. Between 2000 and 2006 alternating erosion and depositional shorelines were formed and pronounced deposition and erosion occurred along the coast of Pattukotai and Tiruvadanai. Between 2010 and 2015 alternating erosion and depositional shorelines were formed (Table 6.2) and peak deposition and erosion were identified along the coast of Tirutturappundi (Figs. 6.6, 6.7, 6.8, 6.9, 6.10 and 6.11).

In Sect. 6.2, between the years 1988 and 1995 alternating erosion and depositional shorelines were formed. The highest deposition and erosion were identified along the coast of Mandabam. Most of the deposition occurred along Devipattinam. Between 1995 and 2001 alternating erosion and depositional shorelines were formed. Highest deposition and erosion were identified along the coast of Uchipuli. During 2000 and 2006 alternating erosion and depositional shorelines were formed with pronounced

Table 6.2 Erosion and accretion rates of shoreline (Sect. 6.1)

Section 1					
Sl. no	Year	Deposition	Place	Erosion	Place
1	1988–1995	11.9	Tirutturaippundi	−11.5	Peravurani
2	1995–2000	12.2	Mannargudi	−11.0	Arantangi
3	2000–2006	10.9	Pattukottai	−10.7	Tiruvadanai
4	2006–2010	10.9	Tiruvadanai	−11.8	Mannargudi
5	2010–2015	10.9	Tirutturaippundi	−10.9	Tirutturaippundi
6	1988–2015	8.4	Mannargudi	−5.4	Tirutturaippundi

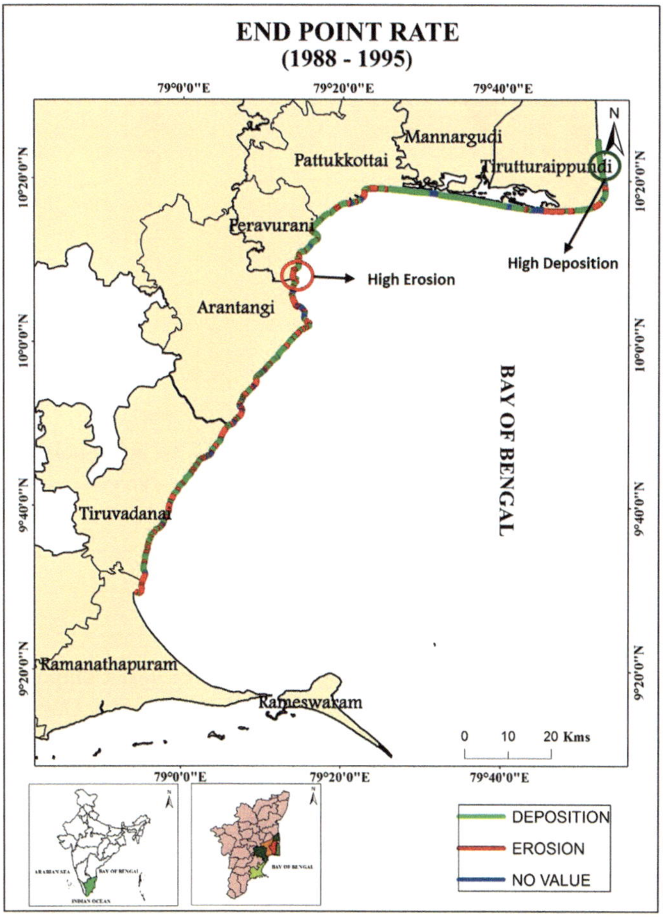

Fig. 6.6 EPR (1988–1995)

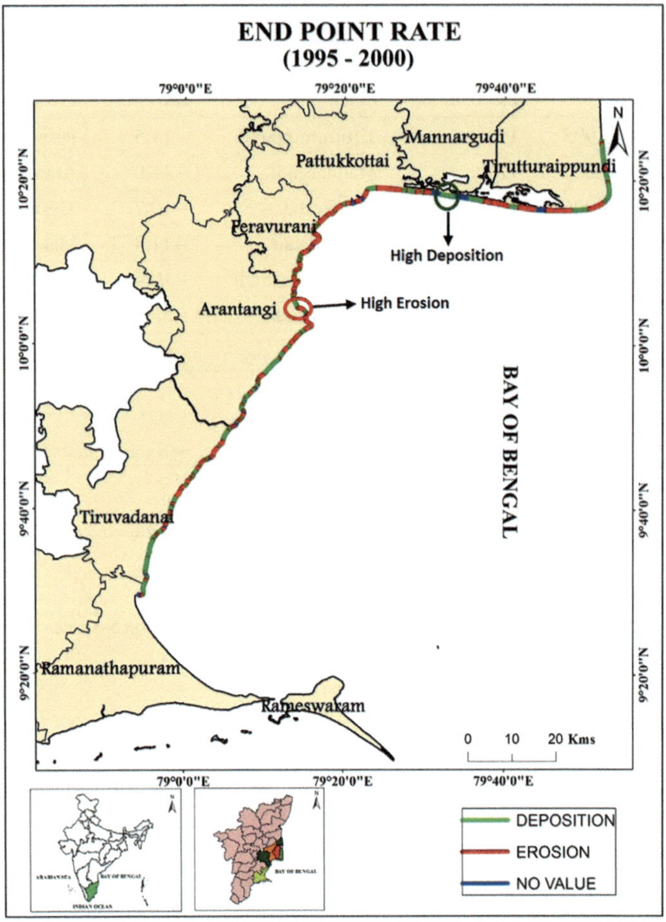

Fig. 6.7 EPR (1995–2000)

deposition and erosion activity registering along the coast of Uchipuli and Mandabam. Between 2010–2015 (Table 6.3), erosion and deposition at Devipattinam and Mandabam accrued (Figs. 6.12, 6.13, 6.14, 6.15, 6.16 and 6.17).

6.4.2 *Assessment of Shoreline Change Along Coastal Structures*

Shoreline change along coastal structures was estimated from Nagapattinam to Ramanathapuram district wherein a total of 19 coastal structures classified into 13 sites. The structures were mostly dockyards and a few jetties. The span of shoreline

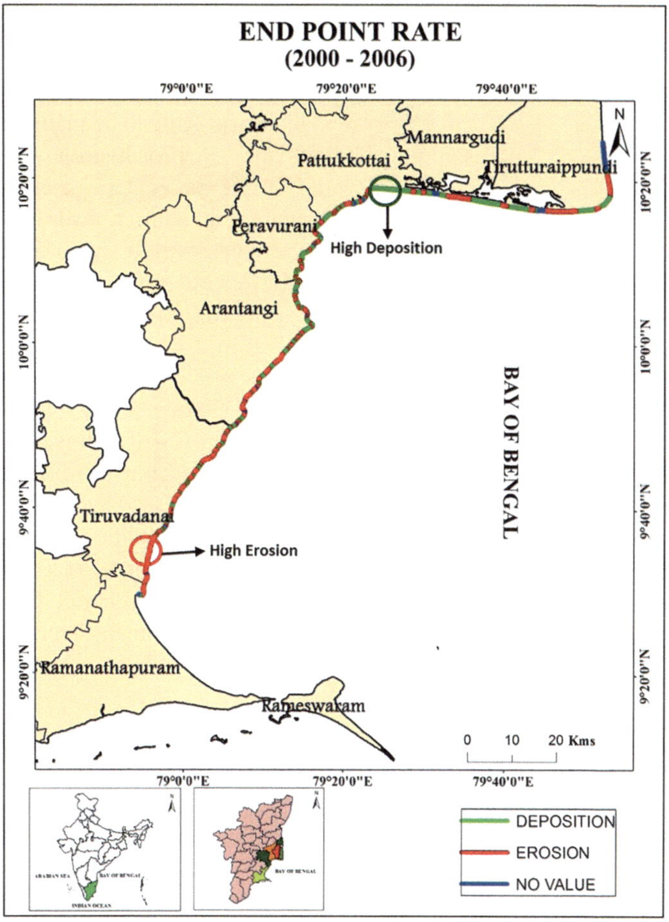

Fig. 6.8 EPR (2000–2006)

length taken into consideration varied for each site (Natesan et al. 2014; Raj et al. 2019; Rangel-buitrago et al. 2017) (Table 6.4).

Overall change in the shoreline at Site 2 deposition was observed on the southern side of the structure and both erosion and deposition were observed on the eastern side of the structure between the years 2010 and 2016. At Site 3 overall change shows deposition on both sides of the structure for the years 2013 and 2016. At Site 4 deposition was observed along the western side of the structure and some erosion was observed along the eastern side of the structure for the years 2013 and 2016. At Site 5 the overall change in shoreline shows deposition along the southern side and erosion along the northern side of the structure during 2008–2016. At Sites 6 and 7 there was mostly deposition over the years 2008 and 2016, though nominal erosion also occurred at Site 7 on the eastern side of the jetty between 2004 and 2016. At

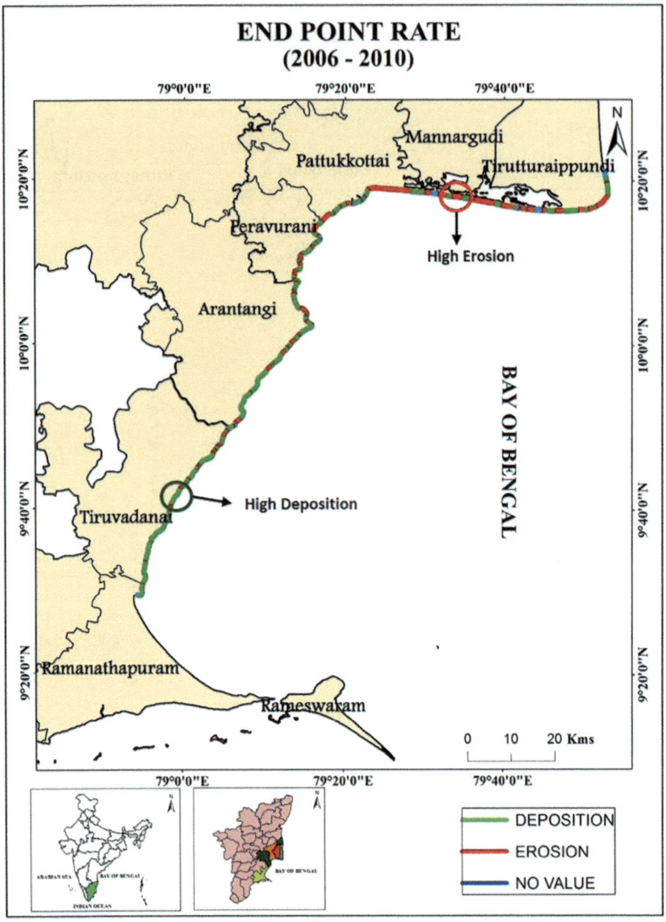

Fig. 6.9 EPR (2006–2010)

Site 8 most of the depositional features were found along the southern side of the structure and at the northern side alternated with deposition along with erosion during 2004–16. Site 9 saw both erosional and depositional features accruing, with more deposition than erosion. At Site 10 mostly erosion was observed along either side of the jetty. Small amount of deposition was observed along the southern side of the jetty during 2002–16. At Site 11, erosional features were observed along the southern side of the structure and along the northern side of the structure, both erosional and depositional features were observed. The predominant feature was erosion for the years 2003–2016. At Site 13, along the southern side of the structure, both erosional and depositional features were observed and along the northern side it was mostly erosion and a few depositional features observed close to the structure. At Site 14, erosional features were observed along both sides of the structure during the years

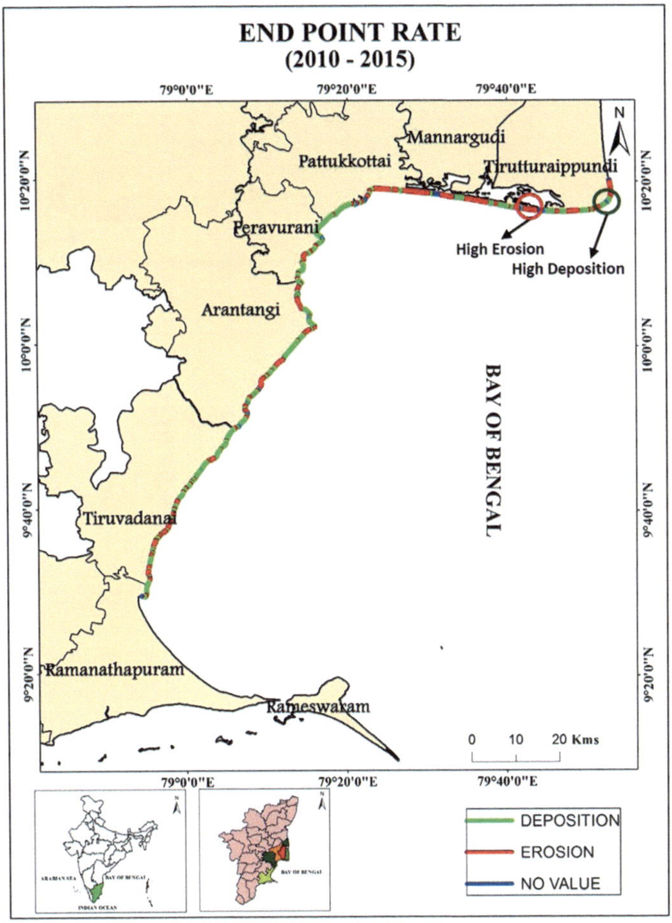

Fig. 6.10 EPR (2010–2015)

2003–2016 (Figs. 6.18, 6.19, 6.20, 6.21, 6.22, 6.23, 6.24, 6.25, 6.26, 6.27, 6.28 and 6.29).

6.5 Conclusions

The study sought to analyse long term changes in shoreline during the years 1992 to 2015 and the impacts of the construction of hard structures along the coast of Nagapattinam to Ramanathapuram. Shorelines were demarcated using LANDSAT imageries. The entire coastline was segmented into 100 m transects. Shoreline accretion and erosion rate were calculated at a five-year interval between 1992–2015. The

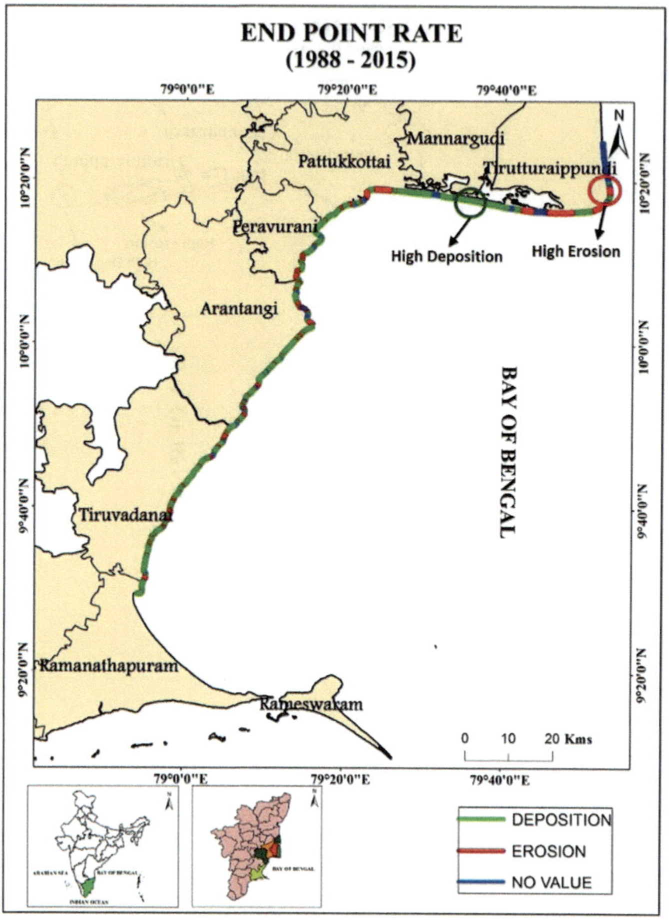

Fig. 6.11 EPR (1988–2015)

Table 6.3 Erosion and accretion rates of shoreline (Sect. 6.2)

Section 6.2					
Sl. no	Year	Deposition	Place	Erosion	Place
7	1988–1995	6.9	Mandabam	−12.3	Mandabam
8	1995–2000	6.4	Mandabam	−10.4	Mandabam
9	2000–2006	9.8	Mandabam	−11	Mandabam
10	2006–2010	6.3	Mandabam	−11.5	Mandabam
11	2010–2015	8.3	Devipattinam	−7.5	Mandabam
12	1988–2015	4.4	Mandabam	−6.6	Mandabam

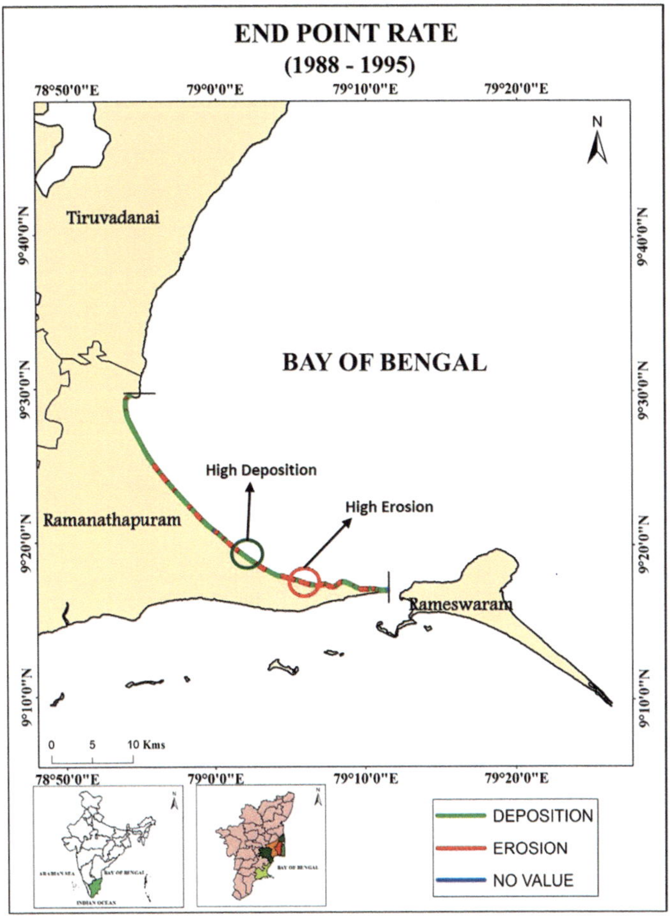

Fig. 6.12 EPR (1988–1995)

end results were then tabulated and maps were prepared for every output demarcating the highest rates of erosion and deposition. To determine the impacts of hard coastal structures a total of 18 structures along 14 sites were examined. Coastal structures and shorelines were demarcated from Google Earth imageries. The following conclusions were drawn from the present work:

1. Along the coast of Nagapattinam to Ramanathapuram depositional were observed prominently in Mannargudi, Arantangi, Tiruvadanai and Ramanathapuram taluk. The highest amount of deposition (upto 12.17 m) was observed in Mannargudi taluk. Erosional features were observed all along the coast in small magnitudes and tended to be higher in Tirutturappundi, Pattukotai and Ramanathapuram taluk. The highest amount of erosion upto (−12.3 m) was observed in Ramanathapuram taluk.

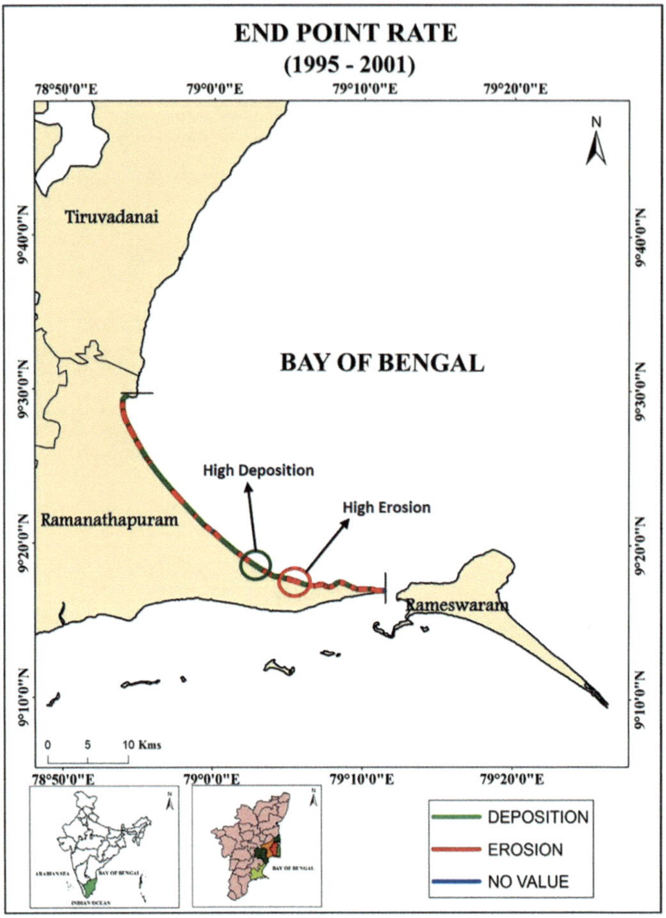

Fig. 6.13 EPR (1995–2000)

2. For the structures present in Rameshwaram taluk, depositional features were observed prominently along all the structures. In some structures, erosion was observed along the eastern side. For Ramanathapuram taluk, depositional features were observed prominently along the structures and erosion in some structures along the northern part. In Arantangi, erosional features were observed prominently along the structures and depositional features on the northern side. In Pattukottai, erosional features were prominent along the structures along with nominal deposition along the southern side.

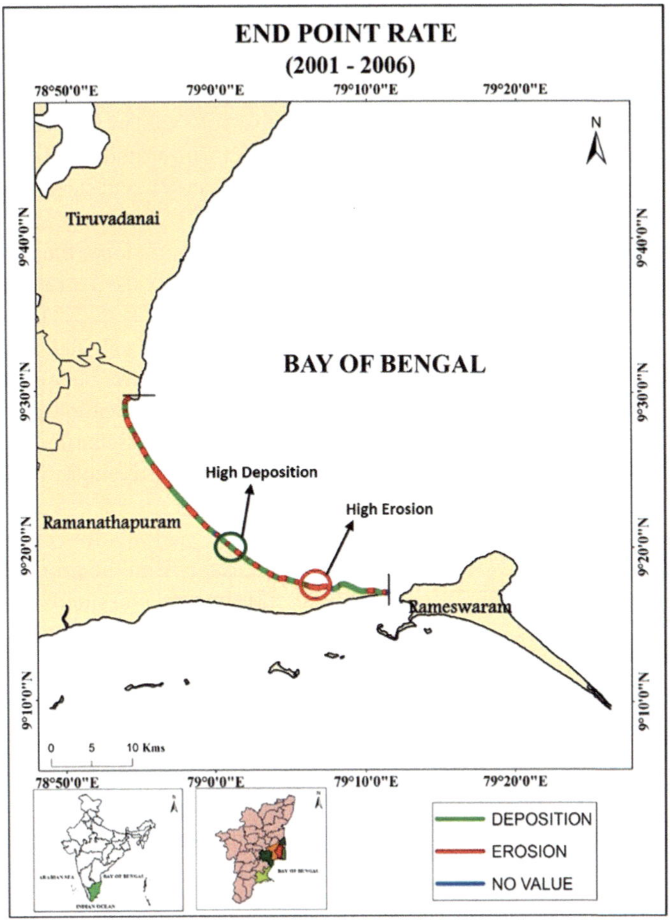

Fig. 6.14 EPR (2000–2006)

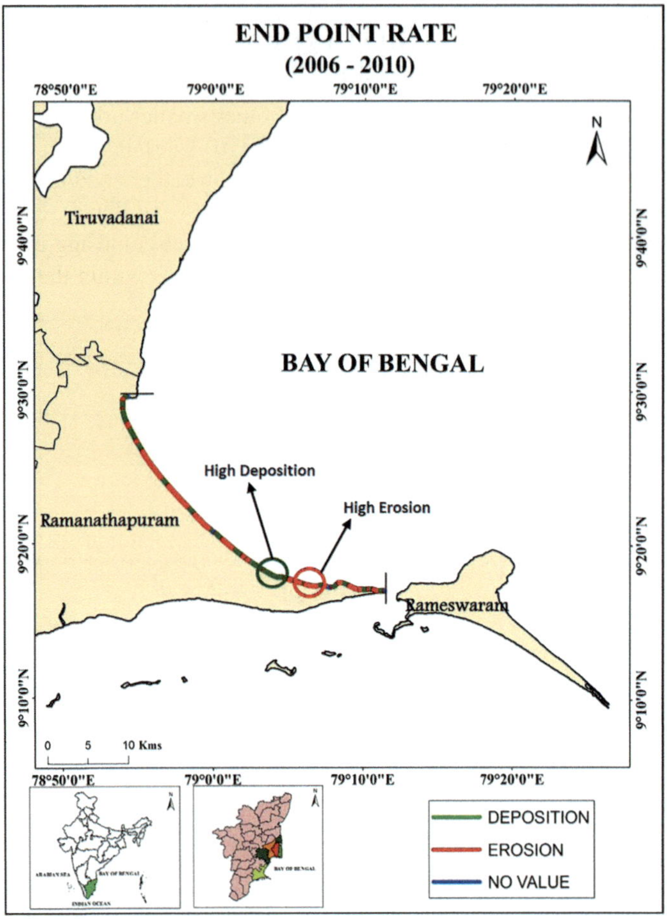

Fig. 6.15 EPR (2006–2010)

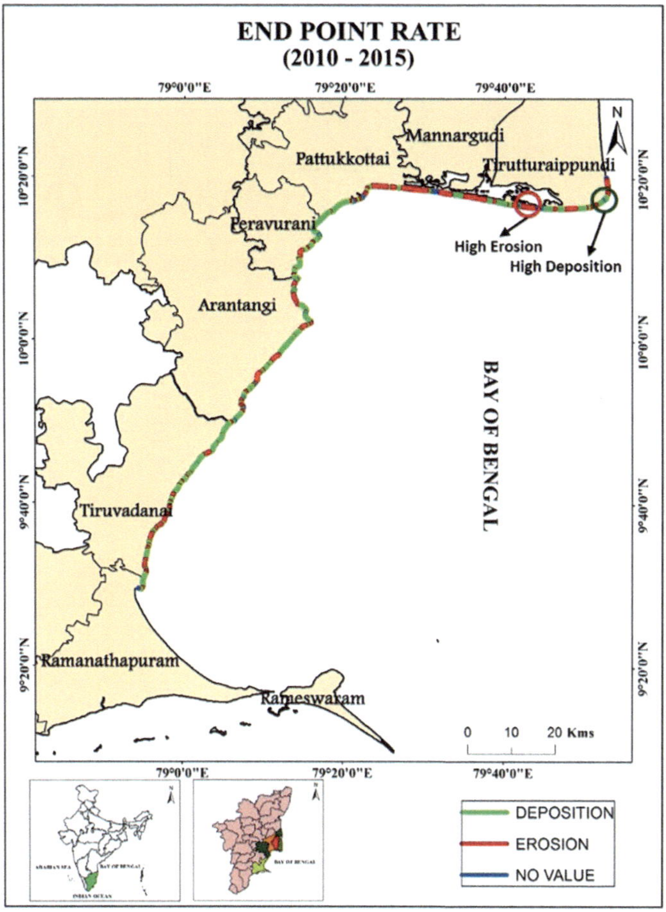

Fig. 6.16 EPR (2010–2015)

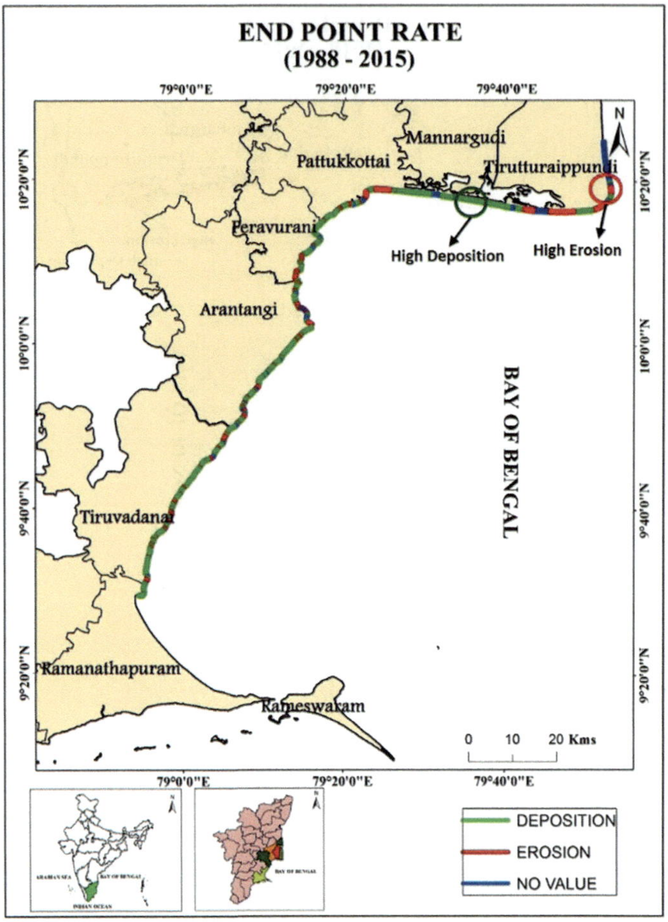

Fig. 6.17 EPR (1988–2015)

Table 6.4 Highest erosion and deposition (NSM) and (EPR)

Site	NSM		EPR	
	Erosion	Deposition	Erosion	Deposition
1	−15.7	16.7	−36.7	28.6
2	−24.2	22.3	−57.6	53.02
3	−6.2	31.9	−10.3	17.7
4	−9.7	14.1	−16	16.2
5	−8.9	17.8	−4.06	6.3
6	−8.7	16.06	−2.3	6.3
7	−7.3	12.3	−3.8	6.09
8	−17.8	11.3	−10	19.6
9	−12.5	11.1	−8.2	13.7
10	−12.5	10.5	−6.6	8.3
11	−18.2	14.6	−11.3	3.1
12	−5.6	12.8	−11.2	7.4
13	−5.5	14.8	−6.5	5.4

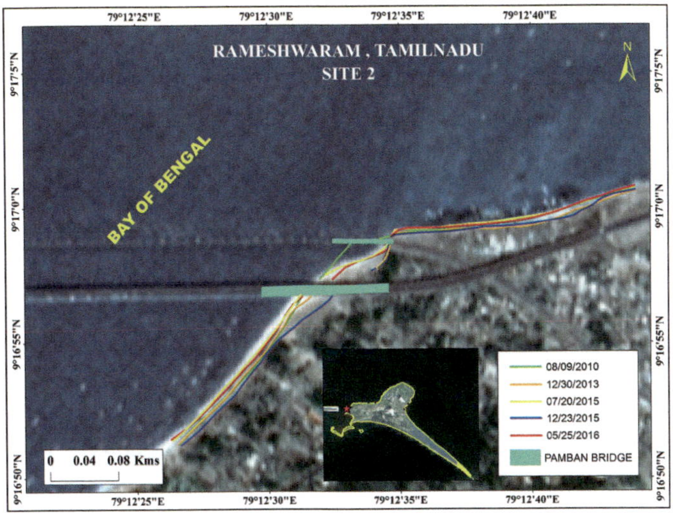

Fig. 6.18 Site 2

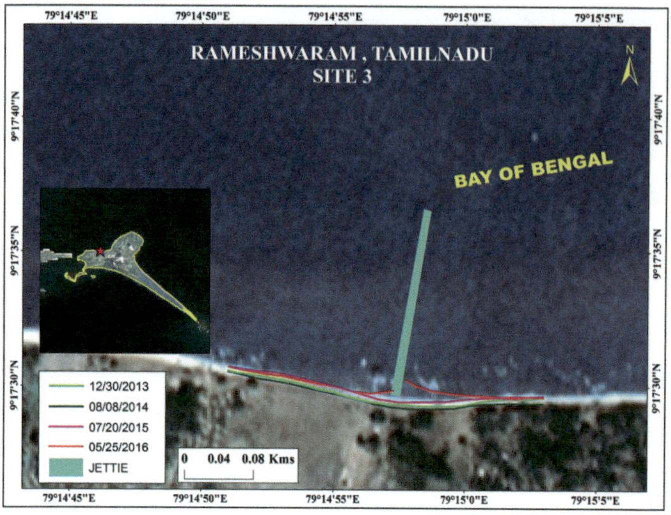

Fig. 6.19 Site 3

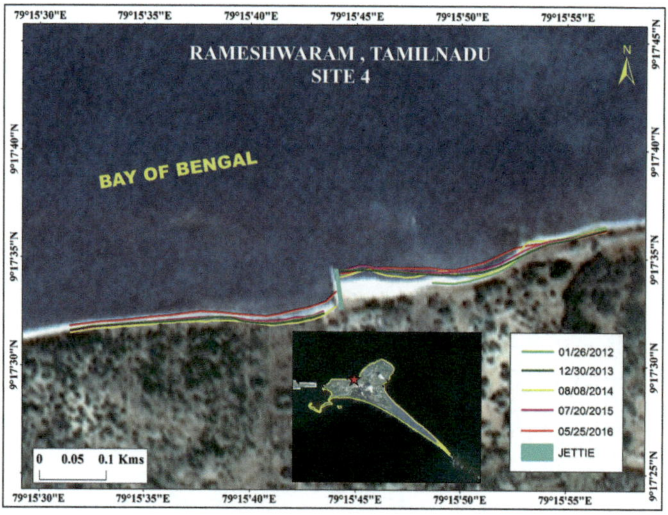

Fig. 6.20 Site 4

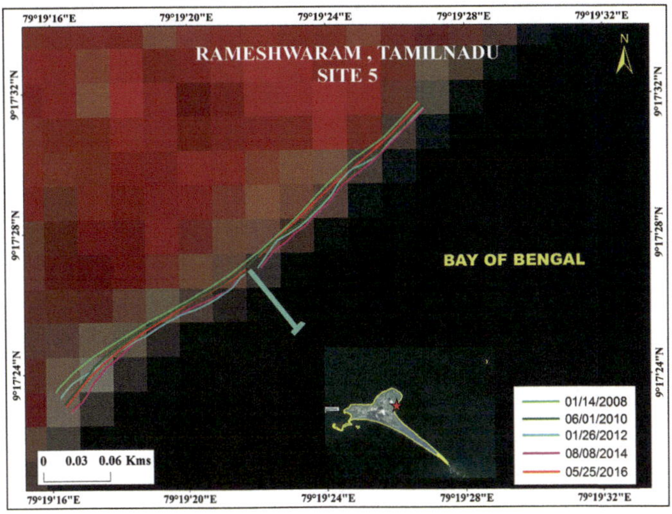

Fig. 6.21 Site 5

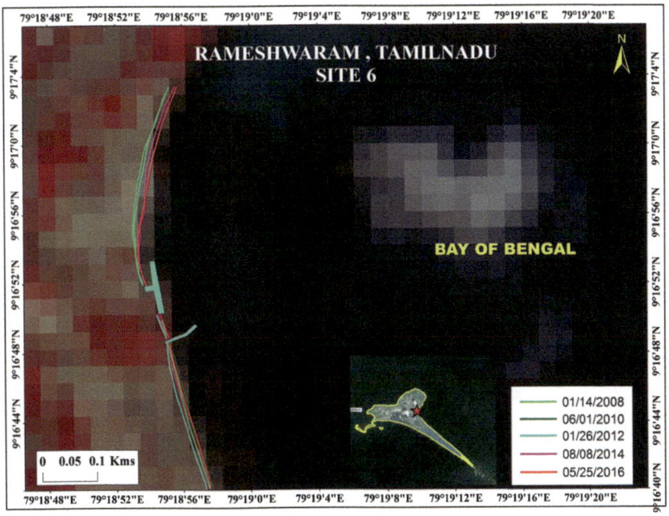

Fig. 6.22 Site 6

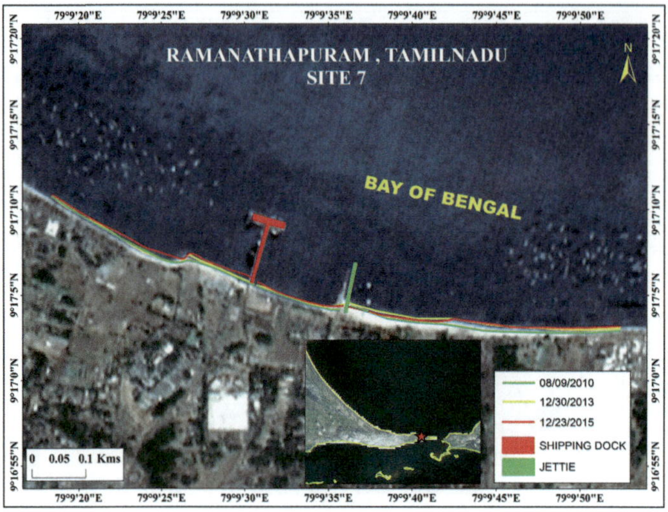

Fig. 6.23 Site 7

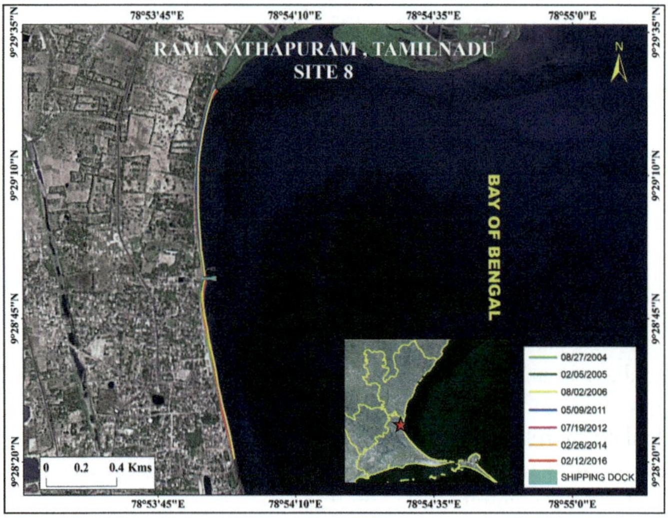

Fig. 6.24 Site 8

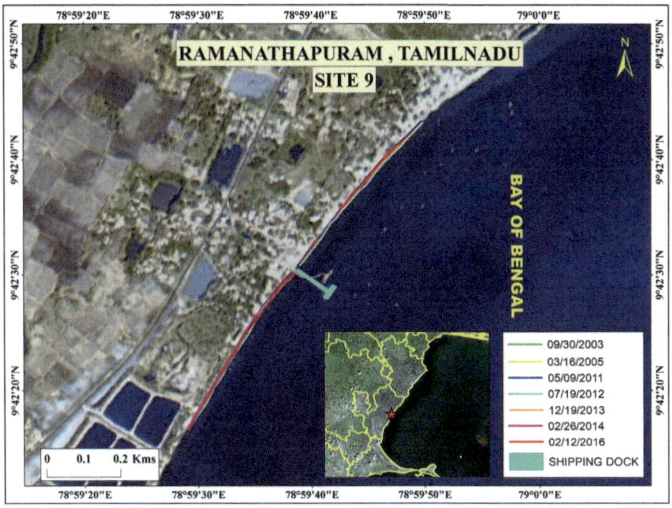

Fig. 6.25 Site 9

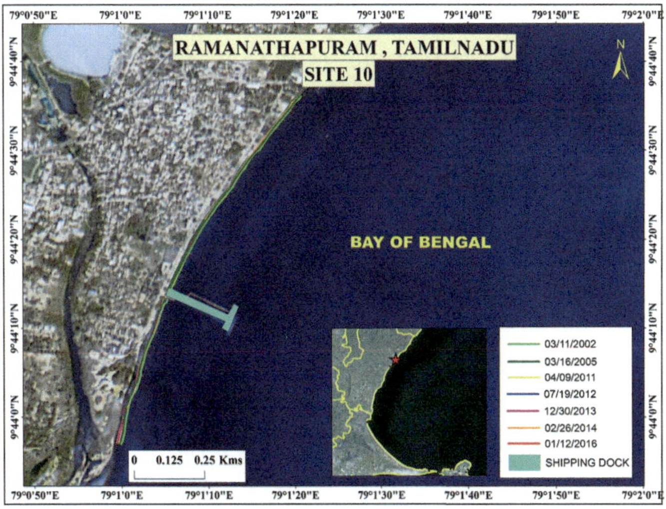

Fig. 6.26 Site 10

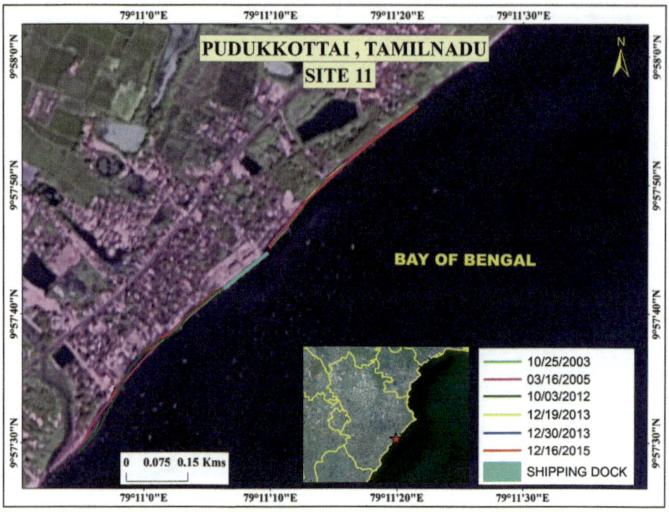

Fig. 6.27 Site 11

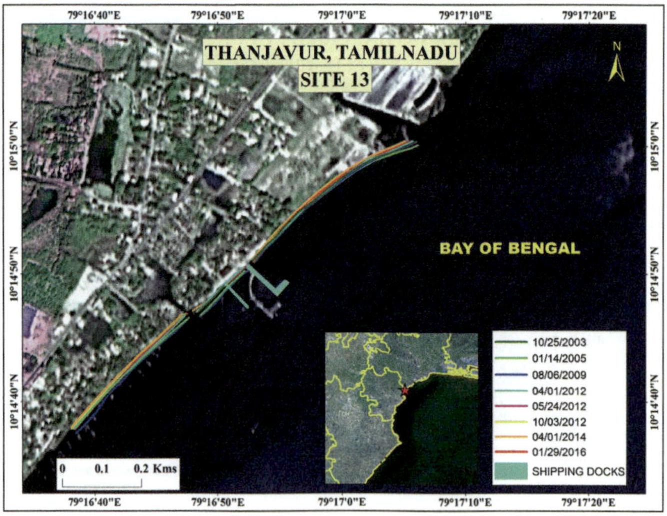

Fig. 6.28 Site 13

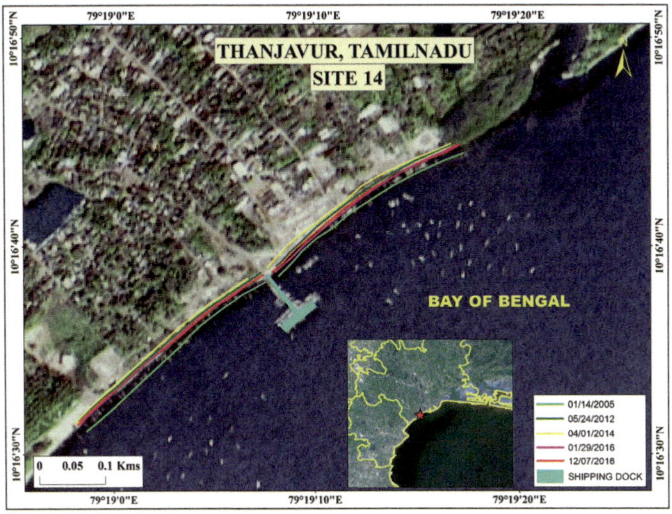

Fig. 6.29 Site 14

Acknowledgements We would like to extend our thanks to Dr. Basant Kumar Jena (Scientist—G) and Dr. Vijaya Ravichandran (Scientist—G), Coastal Environmental Engineering division (CEE), National Institute of Ocean Technology, Chennai, India for their valuable support and time for the research work.

References

Dwarakish GS, Nithyapriya B (2016) Application of soft computing techniques in coastal study—a review. J Ocean Eng Sci 1(4):247–255. https://doi.org/10.1016/j.joes.2016.06.004

Kaliraj S, Chandrasekar N, Ramachandran KK (2017) The Egyptian Journal of Remote Sensing and Space Sciences Mapping of coastal landforms and volumetric change analysis in the south west coast of Kanyakumari, South India using remote sensing and GIS techniques. Egypt J Remote Sens Space Sci 20(2):265–282. https://doi.org/10.1016/j.ejrs.2016.12.006

Kankara RS, Selvan SC, Markose VJ, Rajan B, Arockiaraj S (2015) Estimation of long and short term shoreline changes along Andhra Pradesh coast using remote sensing and GIS techniques. Procedia Eng 116(Apac):855–862. https://doi.org/10.1016/j.proeng.2015.08.374

Kumar A, Narayana AC, Jayappa KS (2010) Geomorphology Shoreline changes and morphology of spits along southern Karnataka, west coast of India: a remote sensing and statistics-based approach. Geomorphology 120(3–4):133–152. https://doi.org/10.1016/j.geomorph.2010.02.023

Lakshmanna B, Prasad TL (2018) Author's accepted manuscript. Society and Environment, Remote Sensing Applications. https://doi.org/10.1016/j.rsase.2018.04.003

Mahendra RS, Mohanty PC, Bisoyi H, Kumar TS, Nayak S (2011) Ocean & coastal management assessment and management of coastal multi-hazard vulnerability along the Cuddalore e Villupuram, east coast of India using geospatial techniques. Ocean Coast Manag 54(4):302–311. https://doi.org/10.1016/j.ocecoaman.2010.12.008

Maiti S, Bhattacharya AK (2009) Shoreline change analysis and its application to prediction: a remote sensing and statistics based approach. Mar Geol 257(1–4):11–23. https://doi.org/10.1016/j.margeo.2008.10.006

Natesan U, Parthasarathy A, Vishnunath R, Kumar GEJ, Ferrer A (2015) Monitoring longterm shoreline changes along Tamil Nadu, India using geospatial techniques. Aquatic Procedia 4(Icwrcoe):325–332. https://doi.org/10.1016/j.aqpro.2015.02.044

Natesan U, Rajalakshmi PR, Ferrer VA (2014) Shoreline dynamics and littoral transport around the tidal inlet at Pulicat, southeast coast of India. Cont Shelf Res 80:49–56. https://doi.org/10.1016/j.csr.2014.02.018

Raj N, Gurugnanam B, Sudhakar V, Francis PG (2019) Geodesy and Geodynamics Estuarine shoreline change analysis along the Ennore river mouth, south east coast of India, using digital shoreline analysis system. Geodesy Geodyn 10(3):205–212. https://doi.org/10.1016/j.geog.2019.04.002

Rangel-buitrago N, Williams A, Anfuso G (2017) Ocean & Coastal Management Hard protection structures as a principal coastal erosion management strategy along the Caribbean coast of Colombia. A chronicle of pitfalls. Ocean Coastal Manag. https://doi.org/10.1016/j.ocecoaman.2017.04.006

Saleem A, Awange JL (2019) Catena Coastline shift analysis in data deficient regions: exploiting the high spatio-temporal resolution Sentinel-2 products. CATENA 179(March):6–19. https://doi.org/10.1016/j.catena.2019.03.023

Sankari TS, Chandramouli AR, Gokul K, Surya SSM, Saravanavel J (2015) Coastal vulnerability mapping using geospatial technologies incuddalore-pichavaram coastal tract, Tamil Nadu, India. Aquatic Procedia 4(Icwrcoe):412–418. https://doi.org/10.1016/j.aqpro.2015.02.055

Saravanan S, Parthasarathy KSS, Vishnuprasath SR (2019) Monitoring spatial and temporal scales of shoreline changes in the Cuddalore. Elsevier Inc., Coastal Zone Management. https://doi.org/10.1016/B978-0-12-814350-6.00004-5

Saxena S, Geethalakshmi V, Lakshmanan A (2013) Development of habitation vulnerability assessment framework for coastal hazards: Cuddalore coast in Tamil Nadu, India—a case study. Weather Climate Extremes 2:48–57. https://doi.org/10.1016/j.wace.2013.10.001

Sowmya K, Sri MD, Bhaskar AS, Jayappa KS (2019) Long-term coastal erosion assessment along the coast of Karnataka, west coast of India. Int J Sedim Res 34(4):335–344. https://doi.org/10.1016/j.ijsrc.2018.12.007

Sreenivasulu G, Jayaraju N, Raja BCS, Prasad TL, Lakshmanna B, Nagalakshmi K, Prashanth M (2016) ScienceDirect River mouth dynamics of Swarnamukhi estuary, Nellore coast, southeast coast of India. Geodesy Geodyn 7(6):387–395. https://doi.org/10.1016/j.geog.2016.09.003

Vivek G, Goswami S, Samal RN, Choudhury SB (2019) Data in brief monitoring of Chilika lake mouth dynamics and quantifying rate of shoreline change using 30 m multi-temporal Landsat data. Data Brief 22:595–600. https://doi.org/10.1016/j.dib.2018.12.082

Chapter 7
Application of GIS Techniques and Environmental Flow Norms to Assess Hydropower Impact on the Environment: A Case Study of Binwa and Larji Hydropower Projects in Himachal Pradesh, India

C. Prakasam and R. Saravanan

Abstract The Beas is a river in Himachal Pradesh, India which possesses substantial hydropower potentials of about 6000 Megawatts. About 18 hydropower projects targeted to generate about 3270 Megawatts are under construction. Environmental flow, currently gaining attention all over the world is perhaps the most important criterion that needs to be maintained for hydropower generation. This study is aimed at identifying a suitable location for the construction of a dam using weighted overlay analysis and to identify appropriate environmental strategies for hydropower generation. The analysis is based on terrain characterization such as elevation, slope, stream order, stream number, and other related indices. A land use land cover (LULC) map and precipitation data are additional variables used in the analysis. A site suitability map and location of a potential hydropower project were generated using weighted overlay method in a GIS environment. The map is also compared with the currently proposed location maps for validation. Environmental flow guidelines that are currently adopted in numerous countries around the world were reviewed to evaluate and recommend the best strategy for the Beas river basin. This outcome of the study provides recommendations for better planning and management of the environment and ecosystem of the Beas river basin. Such analysis and planning strategies can be applied to other river systems of similar kinds of terrain as well.

Keywords GIS · Environmental flow · Potential zones · Weighted overlay method

C. Prakasam (✉)
Department of Geography School of Earth Sciences, Assam University, Diphu Campus (A Central University), Diphu, Karbi Anglong, Assam 782462, India
e-mail: cprakasam@gmail.com; c.prakasam@aus.ac.in

R. Saravanan
Tata Consulting Engineering Limited (TCEL), Bangalore, Karnataka 560001, India

A. Saikia and P. Thapa (eds.), *Environmental Change in South Asia*,
https://doi.org/10.1007/978-3-030-47660-1_7

7.1 Introduction

Hydropower is one of the important widely recognized sustainable sources of green energy such as solar and wind energy. Further, a hydropower plant (HPP) is considered a flawless, inexhaustible, and non-polluting source of energy. The concept of environmental flow, which is of paramount importance for hydropower development, is either not clearly defined and/or yet to be strictly implemented. HPPs possess higher effectiveness in terms of long term management and financial implications in addition to being environmental and a clean source of energy.

Located in Western Himalayas, Himachal Pradesh, a state in northern India, is characterized by perennial river sources, which provide enormous hydropower generating potential. It is estimated that the Beas river basin has a potential capacity to produce upto 20,000 MW of hydroelectricity as the State of India's Rivers, 2016 (https://indiariversforum.org/) noted. The government of Himachal Pradesh installed several projects in the Beas river basin. However, the Government often overlooks the impact of the construction and operation of hydropower project upon the surrounding environment. One of the mitigation measures to reduce the environmental impacts adopted by the Government of India (GoI) is the formulation of the law passed by the National Green Tribunal (NGT) which recommends that "15% of the average lean season flow should be maintained as the environmental flow". Environmental flow is termed as the minimum flow that has to be sustained or released downstream of the hydropower project to maintain the health of the ecosystem. In the study area, the Larji hydropower project is located upstream of the Pandoh dam. To the downstream of the Pandoh dam, the Patikari hydropower dam site has been constructed. Therefore, the point of maintaining the ecological flow is questionable keeping this construction in view. In addition, various other factors contribute significantly to the potential hydropower project site. An issue that is of concern pertains to the amount of environmental flow that would be possible to be maintained.

In this analysis, the delineation of the potential sites for hydropower dam was carried out using the weighted overlay of various parameters determined from the DEM. Several analyzes related to groundwater, aquifer, hydropower, rainwater harvesting, potential zone mapping, and identifying a suitable location for hydropower generation have been examined. Tarife et al. (2017) applied GIS techniques to model suitable hydrological sites in the Philippines using various data such as meteorological, topographical and soil, DEM, LULC, and watershed maps. Their findings revealed about 62% sites were potentially suitable sites for micro-hydroelectric projects (5–100 kW) and 38% sites were favourable for the construction of less than 5 KW hydropower project. Larentis et al. (2010) delineated the potential hydropower sites by analyzing regional streamflow data in GIS-based computational program called Hydrospot. Identifying probable sites for hydropower generation, along the river was found to be more helpful than the traditional survey method of spotting different hydropower sites. Similarly, Sahu and Prasad (2018) analyzed the DEM for the Hasdo Bango hydroelectric project in Korba district, Chhattisgarh, India to create watershed (fill, direction, and accumulation map), including stream network, and hydrological data (runoff). Lakshmi and Sarvani (2018) considered soil type, LULC, topography, contour, rainfall data, and availability of water in the river as significant parameters for site suitablility analysis. Rising demands for energy has also led to increasing demand for constructing small hydropower projects (Goyal et al. 2015). Satellite imageries along with other collateral data can be used to select small hydropower plants and for the identification of suitable locations for water harvesting structures. In the Mandri watershed, Chhattisgarh, India for water resource development such as gabion, gully plug, check dams, boulder check, boribandhan, and stop dams Saha et al. (2018) used GIS and remote sensing datasets and techniques. They followed the Multi-Criteria Decision Analysis (MCDA) and an Analytical Hierarchy Process (AHP). Iftikhar et al. (2016) incorporated geological maps, discharge data, rainfall, Landsat 5, and ASTER Global DEM of Swat valley to evaluate the feasibility of selected locations to overcome energy crisis, water shortage, and natural disasters that reduced the agricultural efficiency in Pakistan. Jasrotia et al. (2019) used the aquifer and terrain characteristics to delineate the artificial recharge zones for the Northern Western Himalayas. They delineated suitable sites using groundwater modeling and geospatial technologies. Various thematic inventories of hydrology, geology, and geomorphological layers using geospatial technologies to delineate the groundwater resources zone, and its validation using the topography data were similarly utilized by other researchers (Haque et al. 2020; Khan et al. 2020). Remote sensing techniques have also been adopted as an efficient and cost-effective survey for groundwater zonation mapping. Gnanachandrasamy et al. (2018) used thematic layers such as drainage, geology, drainage density, slope, and rainfall and adopted the inverse distance weightage (IDW) method to identify groundwater potential zones. There exists considerable literature to support the fact that recent advancements in GIS and remote sensing techniques has facilitated the identification of hydropower project sites in a time bound and cost-effective manner. Such techniques have enhanced our capability to explore water resources (Wang et al. 2014), creating zonation maps, which facilitate decision-making and planning of regional renewable energy by incorporating data on regional potentials and restrictions to various stallholders (Singh et al. 2017; Sarkar et al. 2020). GIS offers the advantage of weighted overlay analysis for the user to specify the relative weights as percentages, decimals, or relative weightings (Samanta and Aiau 2015). Detection of the run of river plant potential location by coupling the hydrological model and GIS techniques (Sammartano et al. 2019), can effectively improve the process of delineating the hydropower sites. Studies conducted by Jena et al. (2020) indicated the density of lineament drainage as being the least sensitive to the potential zone identification, while soil and LULC had higher sensitivity to the groundwater potential zonation. Nithya et al. (2019) confirm that GIS-based AHP method is effective for the identification of groundwater potential zone for sustainable groundwater planning and management. Collection of raw data; processing the raw data into various thematic layers, and modeling the thematic maps are the three main processes involved in the potential zoning of groundwater sources (Al-Ruzouq et al. 2019). Remote sensing and GIS techniques have also been applied in different decision-making strategies, using fuzzy logic, AHP, weighted overlay analysis, and artificial intelligence (AI) for the flexibility of incorporating spatial information (Kumar et al., 2008) and to reclassify and standardize multiple thematic layers using the Jenks technique (Chen et al. 2013). Hassanet al. (2020) integrated the AHP with GIS technique and weighted overlay analysis to assess land suitability analysis in India's Jammu and Kashmir regions.

This Chapter reviews the various minimal flow norms and regulations followed by various countries including those prevalent in India. The Chapter focuses on two basins of the Beas river in Mandi and Kangra districts in Himachal Pradesh, India. The Beas river basin covers parts of the two Indian states of Punjab and Himachal Pradesh. GIS has been used here to explore HPP site suitability.

7.2 Study Area

The Beas basin covers a catchment area of 20,303 km^2 and an estimated hydropower potential of 5995 MW. It spans over a length of 460 km originating from Rohtang Pass and flows from Larji to Talwara and ultimately drains into the Sutlej at Harike, in Punjab, India. About 9,125 MW of major hydropower projects (>25 MW) and 61.8 MW of small hydropower projects (<25 MW) are currently operational. This includes nine major hydropower projects, which falls within the purview of the Himachal Pradesh State Electricity Board Ltd, Bhakra Beas Management Board, and Punjab State Power Corporation. Additionally, there are about 10 small hydropower projects with less than 25 MW capacity, 18 hydropower projects under construction, and many other projects in the pipeline.

The Larji hydropower project of 126 MW capacity and Binwa hydropower project of 6 MW capacity was chosen as a study area (Fig. 7.1) to represent the Beas river basin. Larji catchment area is spread over an area of 4921 sq. km. The large dam was constructed in 1984 and completed in September 2007. It is located at an altitude of 2299 m above mean sea level (MSL). The project has a design head of 56.84 m, 250 m^3/s discharge capacity, and a live storage capacity of 230-hectare meters. The Binwa watershed is located in Kangra district of Himachal Pradesh, India. It lies between 31° 53′ 15″ to 32° 11′ 58″ N latitude and 76° 34′ 08″ to 76° 45′ 53″ E longitude between the Lesser Himalayas and Siwalik hill slopes. The watershed is an agro-eco region of 340.1 km^2.

7.3 Materials and Methods

The materials and methods to achieve the two objectives i.e., (a) to review the existing norms for the environmental flow is based on close reading and evaluation of existing literature, rules and regulations, and (b) the identification of potential locations for the construction of hydropower projects, is based on the quantitative analysis of remote sensing data.

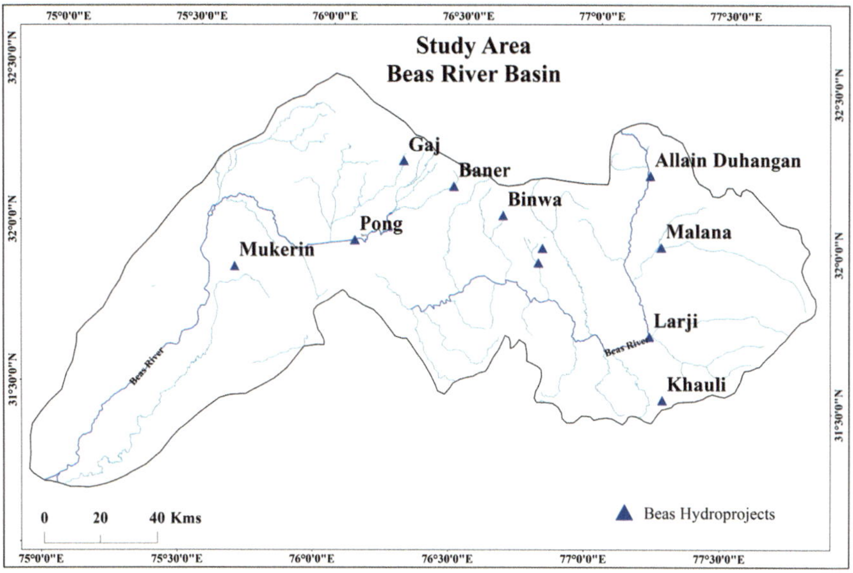

Fig. 7.1 Location map of the various hydropower project in Beas river basin *Source* Beas river basin was delineated from topo sheets and the location of the hydropower projects were obtained from Google Earth

7.3.1 Environmental Flows Law and Guidelines Internationally

The definition of environmental flow and the laws and guidelines to maintain it differs from one country to another. The regulations are expected to provide a practical, legitimate, and sensible action for ecological flows. Some countries quantify the minimum flow, while others use qualitative norms to define the laws and guidelines (Table 7.1).

7.3.2 Minimal Flow Norms in India

India realized the importance of environmental flow in the 1980s and the term "minimum flow" was first coined by the Center Water Commission in 1992. The progress of minimal flow stipulations in India is represented in Table 7.2.

Recently NGT has passed a law, which states that 15% of the average lean season flow should be maintained as the environmental flow. However, the recommendation for environmental flow for different hydro projects (Table 7.3) varies considerably.

Table 7.1 Environmental Flows Law and Guidelines in various countries

Country	References	Environmental flow law and guidelines
European Union (EU)	Acreman and Ferguson (2010) and EU: DIRECTIVE 2000/60/EC (2000)	Water System Directive (WFD) states that the discharge of 7.5 to 35% of must be available for flow regime
France	Souchon and keith (2001)	Legal minimum flow (1/40) ADF (Average daily flow) below dams built before the law
The United Kingdom	Souchon and keith (2001)	The authority can decide the minimal flow requirements based upon the demand and constraints
Germany	Water (Federal water act-WHG), March 2010	Various factors are involved in legalizing a definite minimal flow
Switzerland	Swiss water protection act (http://www.bafu.admin.ch)	The law affirms preservation of aquatic ecosystem's health and sufficient water for livelihood
The United States	Water resources act (1963) (http://www.swfwmd.state.fl.us)	States that the environmental flow will be decided by the state water administration based on their requirements
Australia	Halliday and Robins (2001) and Environmental flow Guidelines	The ecological flow should be maintained to overcome extreme flow events in rare cases
South Africa	Wyk et al. (2006) and South African National Water Act (NWA)	Follows the concept of 'ecological standby' which means that essential supply of water must be available so as to guarantee the river ecosystems
China	De-sheng (2010)	The suggestion is that water will be available to maintain imperative ecosystem resources and limits during ensuing water management operations

Table 7.2 Minimum flow stipulations

Year	Minimum flow stipulations
1999	10 m^3/s as the environmental flow (Yamuna river)
2007–2008	10% of the base flow
2010	20% of ecological flow
2012	The authorities decide on the environmental flow
2017–2018	15–20% of the average lean season flow

Source Jain et al. (2014)

Table 7.3 E-flow Recommendations for various projects

Recommendation	Project	References
Environmental flow of 7 m^3/s	Nathpa Jhakri Hydroelectric Project (1500 Megawatt)	Kumar (2009)
Environmental flow of 4 to 10 m^3/s	RHEP-(Rampur Hydroelectric Project) (412 Megawatt)	DHI (2006)
70% all the time and 30% during the lean period	The Upper Ganges basin (2007)	WWF (Worldwide Fund For Nature) (2012)
50–60% of the total flow	Yamuna River through Delhi	Soni et al. (2013)
Andhiyakore 0.388 m3/s, Sundergarh 0.96 m3/s	Mahanadi River (2009)	Sundaray (2013)
225 m^3/s	Triveni Sangam, Allahabad (2013)	WWF (Worldwide Fund for Nature) (2013)
4.70 cumecs	Kalagarh, Ram Ganga River	Sundaray (2013)
Belur (0.07–9.05) m^3/s, Kollegal (18.24−364.88) m^3/s	Cauvery River (2014)	Durbude et al. (2014)
542 m^3/s during lean period	The Ganges River (2010)	Akter (2010)

7.3.3 *Identification of Potential Location for Hydropower Project*

Delineating the potential zones for the location of hydropower projects is based on three main processes, i.e., (i) collection of raw data, (ii) processing of the raw data, and (iii) analyzing the processed data in the model. ASTERDEM data was downloaded from www.earthdata.nasa.gov/ and Landsat 8 data was downloaded from www.earthexplorer.com/. The spatial resolution of both these data sets is 30 m. Terrain features such as slope and stream order were generated from ASTER dem data. Site suitability analysis was carried out through the following steps:

(a) preparation of spatial data, (b) processing of the data, (c) selection of criteria for providing the weight, (d) overlay analysis, (e) relative weights for the total weight calculation (f) locating the selected sites by mapping. The quantification of the weightage for each of the parameters (Table 7.4) are based on supporting literature, expert advice, and the ability of each parameter to derive the optimal results.

The criteria map definition and their calculation for an allocation of weights is a challenging task. The threshold values define the streams whereby following it decides the stream network density (Imran et al. 2019).

The glacial, fluvial processes and peri-glacial characteristics are climatic sensitive influences that possess interrelationships with topographic parameters such as aspect, altitude, or slope angle, in particular. Additionally, the position of the slope and its angle represents the snow cover on the permafrost. The Beas river flowing in the catchment area was demarcated using the topo sheets (Fig. 7.2). Survey of India (SoI) topographical maps on 1:50,000 scale bearing numbers 52 D/4, 52 D/8, 52 D/12, 52

Table 7.4 Weights for various parameters

Parameter	Class	Weightage	Parameter	Class	Weightage
River	Beas	5	Land use land cover	Forest	3
	Beas River	5		Vegetation	3
	Beas Sutlej	4		Barren land	2
	Pandoh	4		Water	5
	Parvathy	2		Glacier	1
	Sutlej	1	NDVI	(−1)–0.055	5
	Uhl	4		0.056−0.15	4
Slope (Degree)	0−19	1		0.16−0.22	2
	20−35	2		0.23−0.29	2
	36−49	3		0.3−0.54	1
	50−61	4	NDWI	(−0.49)–(−0.26)	1
	62−85	5		(−0.25)–(−0.2)	2
Stream order	7th order	5		(−0.19)–(−0.12)	2
	6th order	5		(−0.11)–(−0.04)	4
	5th order	4		(−0.039)–1	5
	4th order	3			
	3rd order	2			
	2nd order	1			
	1st order	1			

D/15, 53 A/14, 53 A/15, 52 E/3, 52 E/5 were consulted for the analysis. A land use map, which is one of the important influencing factors in deciding the location of the hydropower projects was generated using supervised classification of Landsat 8 data (Fig. 7.3). Slope (Fig. 7.4) as a principal descriptor of form, altitude, and drainage area is an influential factor for fluvial and gravitational processes, and surface roughness, which is an indicator of surface material characteristics (Otto et al. 2018). Slope is reclassified into five categories. The lesser the degree of slope, the higher is the suitability for the location of a hydropower plant. The widely used Strahler's method has been used to derive the stream order from the DEM data (Fig. 7.5). The process involves fill, flow direction, flow accumulation, and drainage calculation using map algebra. Seven classes of stream order representing the river have been derived using this method. Stream segmentation is important for the demarcation of catchment area, and to delineate individual streams, and estimate total discharge. This process is considered in different forms of rate and magnitude, and is complex, as it involves a variety of dynamics (Paudel et al. 2016).

Inverse distance weightage (IDW) is a deterministic interpolation method to allocate the interpolated values for the creation of a surface from the point data by the obtained results of mathematical formulas to surrounding measured values (Earls et al. 2007). Figure 7.6 is derived using the normalized difference vegetation index

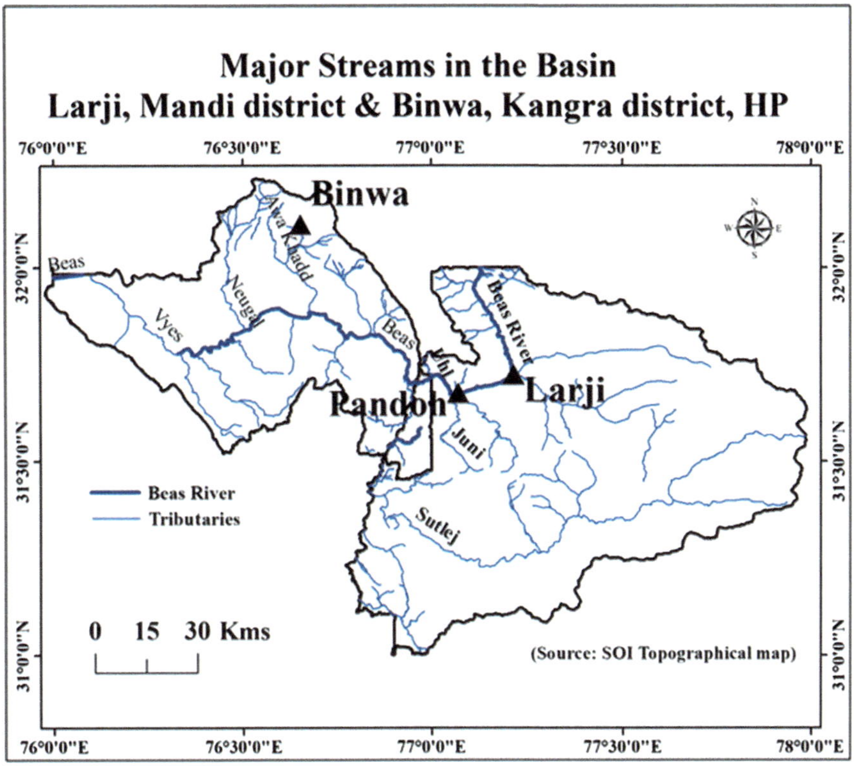

Fig. 7.2 Major streams in the study area

(NDVI), which is a useful method to assess green biomass and in addition to water stress and nutrients (Glenn and Tabb 2019). The Normalized Difference Water Index (NDWI) has been modified by replacement of NDWI used near the infra-red band concerning a middle infra-red band such as LANDSAT TM band 5 (Xu 2006). This index is structured to (a) by green wavelength maximum reflectance of water (b) low water reflectance minimized in NIR, and (c) benefit from the NIR of high vegetation and soil reflectance. As a result, all the water characteristics will have positive values and are enhanced. Therefore, soil and vegetation typically have zero or negative values and are suppressed (Fig. 7.7). The processed thematic layers have been analyzed using the weighted overlay analysis by providing suitable weightage for each parameter (Table 7.4) to delineate the potential location of the hydropower projects.

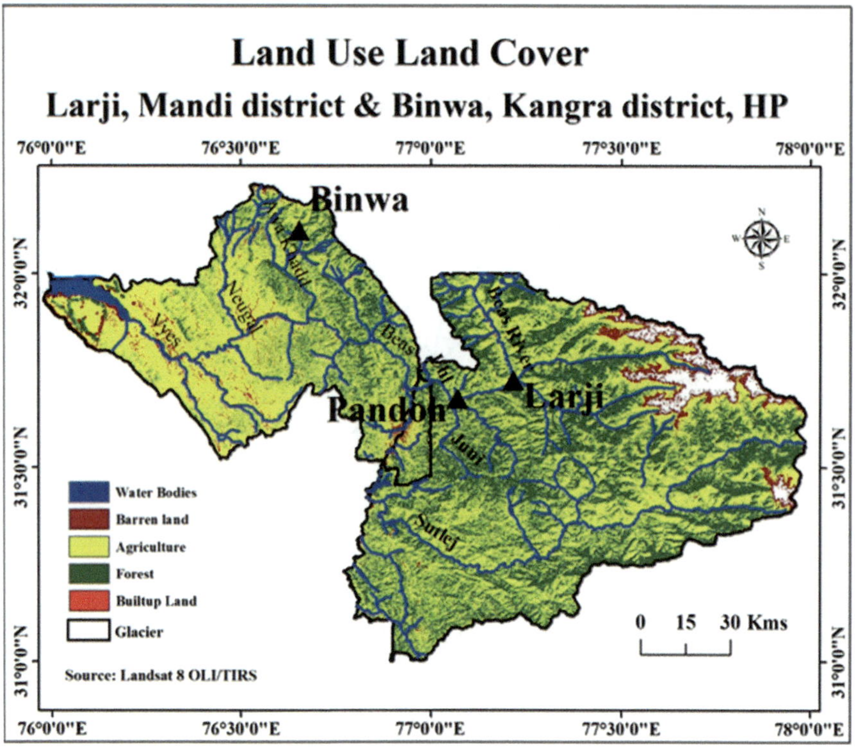

Fig. 7.3 Land use land cover map of the study area

7.4 Results and Discussion

Identifying the potential location of the hydropower project is a challenging process. Yet, the achievement of the objective of this research was made possible through the use of remote sensing and GIS techniques, which provide necessary data, tools and methods for geospatial data processing and site suitability analysis The weighted overlay method can be economical and simulated within a short time period. This method applies to hilly areas or mountainous regions. The weighted overlay with data layers involves classifying a DEM based on elevation data, land use land cover, and stream order and number.

The study area has a moderate slope of 34%. The plain areas with less than 1° slopes are mostly located along the river valley, and are utilized for agricultural purposes. Glaciers prevail in both the basins of over an area of about 31% of the hilly terrain.

The stream order ranges up to the 7th order indicating the volume of water flowing to the river is substantial enough for the location of hydropower projects, provided other factors are favourable. The lower order streams that feed these streams are

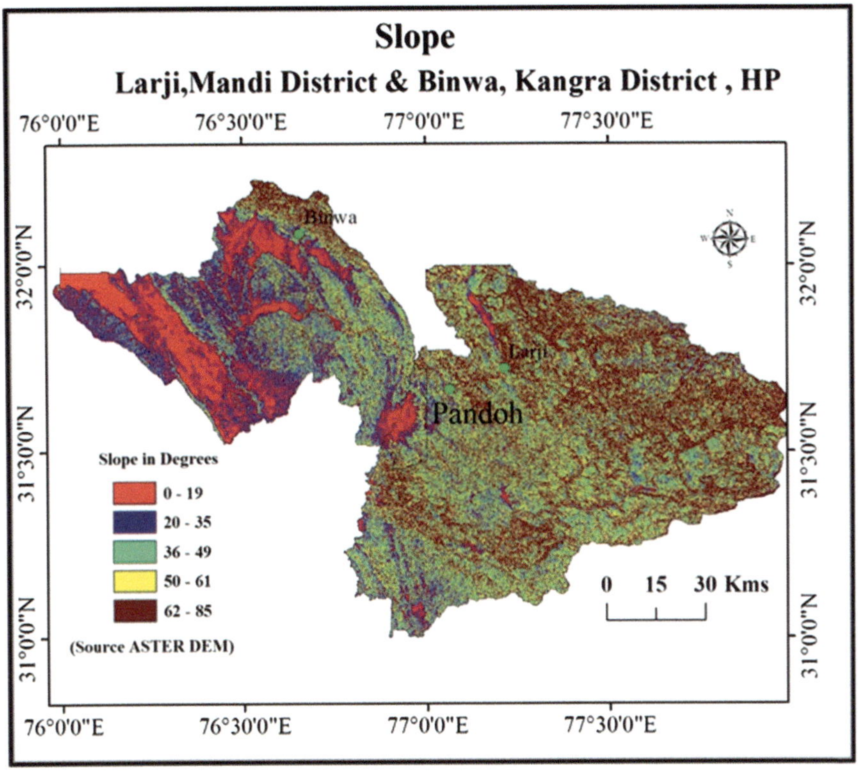

Fig. 7.4 Slope in degrees

mostly rainfed, which ranges from 920 to 2500 mm annually. The intensity of rainfall decreases as we move from Dharamshala to Kullu. The Binwa basin is mostly shrouded in forest and shrubland. The agricultural land area has impacted along the plain region i.e., riverside up to 5.83%. The vegetation index shows that the greenness of the vegetation in the region is average in most of the area and high in the agricultural fields. The assigned weights range from 1 to 9 depending on the importance of the thematic layer and are normalized between 0 and 100 (Table 7.4). This figure represents the areas of potential zones to set up hydropower projects. Based on these, the sites were identified and categorized as possessing low, medium, high, and very high suitability (Fig. 7.8). Very high zones indicate the potential location for the construction of the hydropower project. This location is rich in resources that serve the purpose of locating a hydropower project on a long term sustainable basis. To enable the visualization of the results of the analysis the high suitability potential zones are delineated on a map.

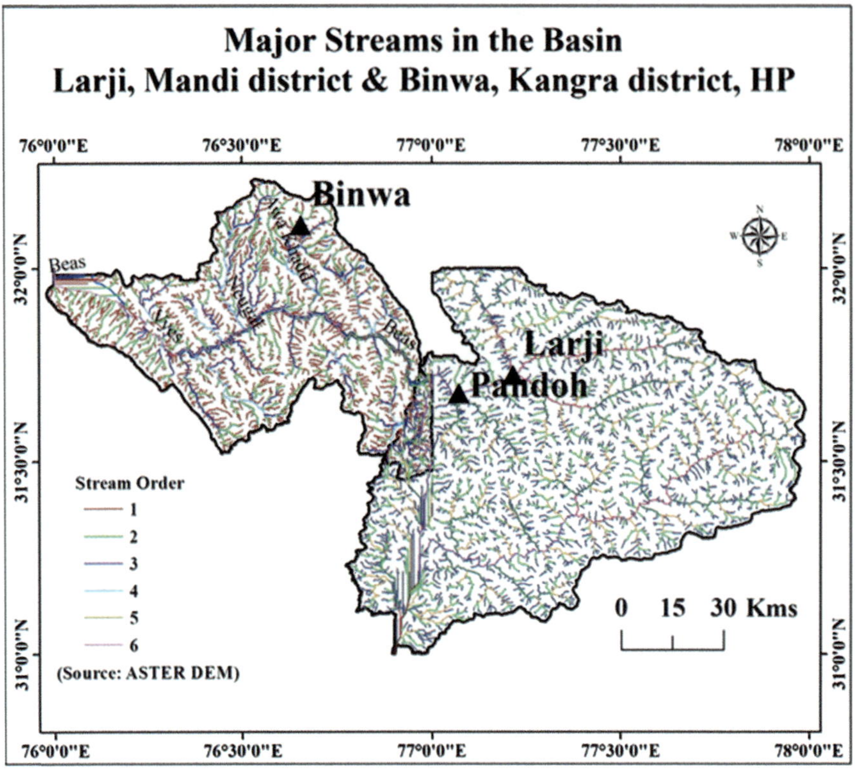

Fig. 7.5 Stream orders based on Strahler's Method

7.5 Discussion

Mountainous regions provide the required geographical conditions for the localization of hydropower projects. Mountain communities rely heavily on the natural resources and livelihood opportunities offered by their habitat. Unfortunately, the construction phase of hydro projects affects the environment due to deforestation, exploitation of land, and other resources upon which the mountain communities depend for their livelihood. Likewise, the operation of hydropower projects also affects mountain communities who primarily depend upon primary sector livelihood activities.

Irrespective of the minimal flow law passed, government, private and individual people must work hand in hand to preserve the environment. Recently the NGT has passed the law of 15 to 20% of average lean season flow as the minimum environmental flow to be maintained in the downstream tracts of the river. The construction of hydropower projects in the identified locations must adhere to such norms so as to be minimally environmentally deleterious.

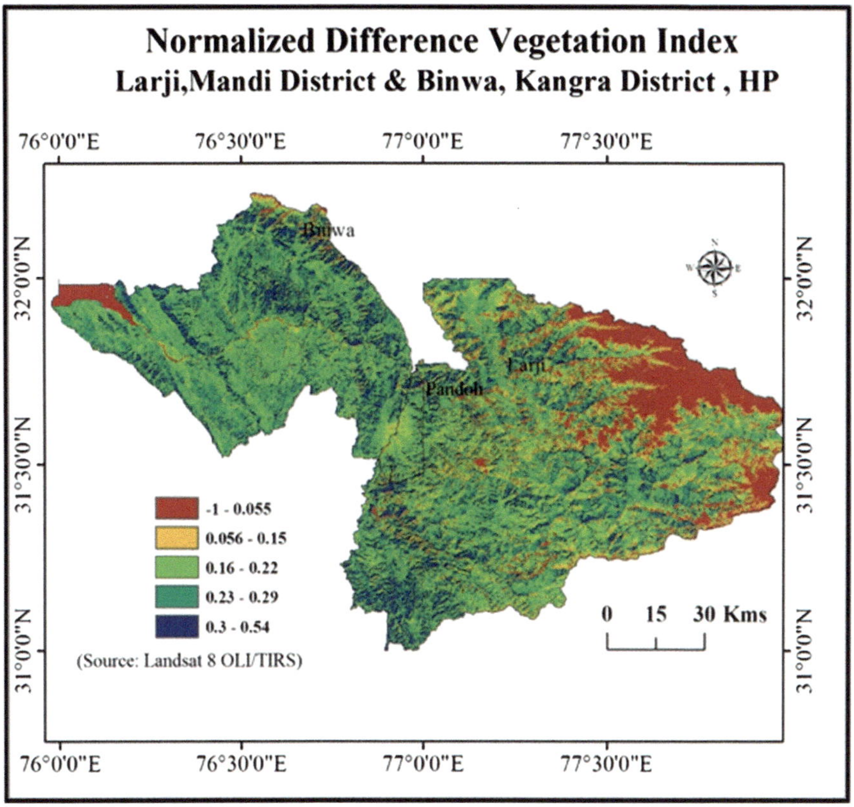

Fig. 7.6 NDVI of the study area

7.6 Conclusion

The construction and operation of the hydropower projects will incur potential impacts on the environment. The framework for a sustainable environment should be practiced by the authorities concerned. Hydropower projects require a large amount of water to generate electricity, hence the amount of water released in the river might be insufficient. As a mitigation measure, the NGT has passed a law that 15% of the average lean season flow should be maintained/ released into the river by the hydropower projects. Small hydropower projects were exempted from this rule. Considering a hypothetical situation where three to four small hydropower projects are being constructed in the place of a large hydropower project, then the environmental flow maintenance could be by passed or avoided. The location for the construction of the hydropower projects wherein the impact would be minimal should also be delineated by governmental authorities. Although suitable precautionary measures and assessments are employed to select sites for the construction of the existing hydropower projects using traditional survey methods, the incorporation

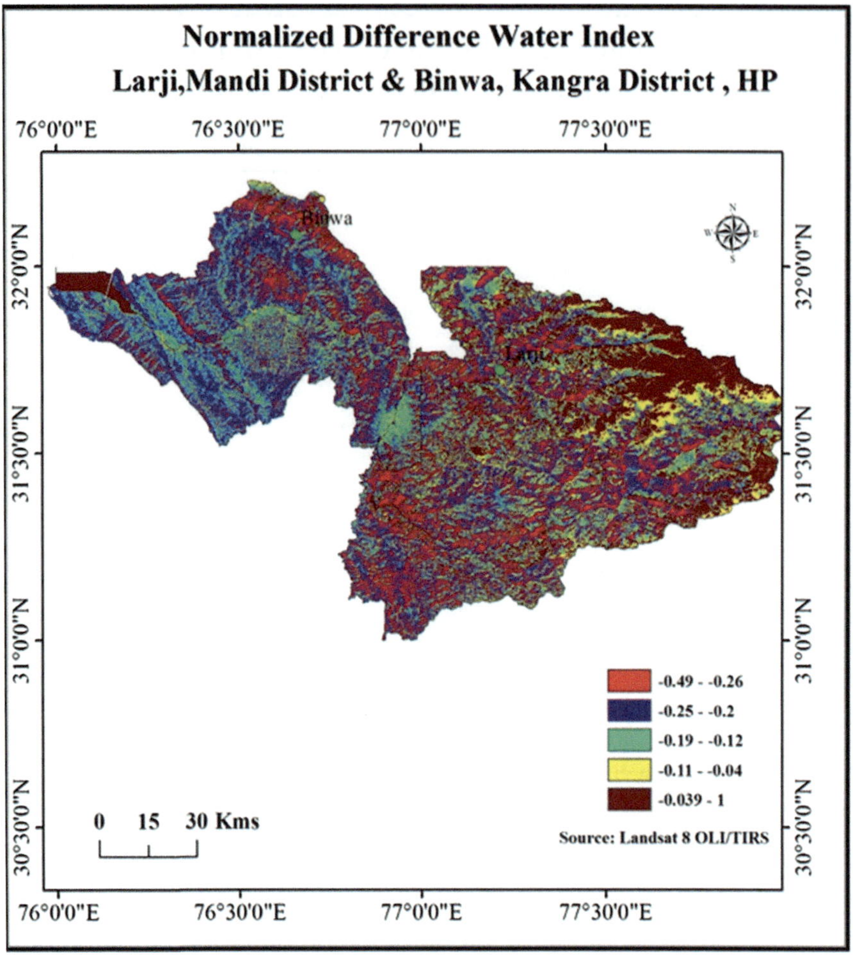

Fig. 7.7 NDWI representing water index of the study area

of remote sensing and GIS techniques to delineate potential zones using a weighted overlay method can prove efficient and cost and time effective. The weighted overlay methodology employed to identify the potential zone for the hydropower project in the present study made it possible to incorporate various parameters, which tend to get neglected when using traditional methods. The criterion for using various layers is primarily established on the watershed characteristics influencing a river's runoff. These characteristics are slope, drainage, NDWI, NDVI, LULC, stream order, etc. Various geospatial tools in GIS were used to create thematic layers as an input to identify the potential locations in the Larji and Binwa basins in the Kullu and Kangra districts of Himachal Pradesh, India. However, field work is always necessary to validate the results from a GIS-based analysis.

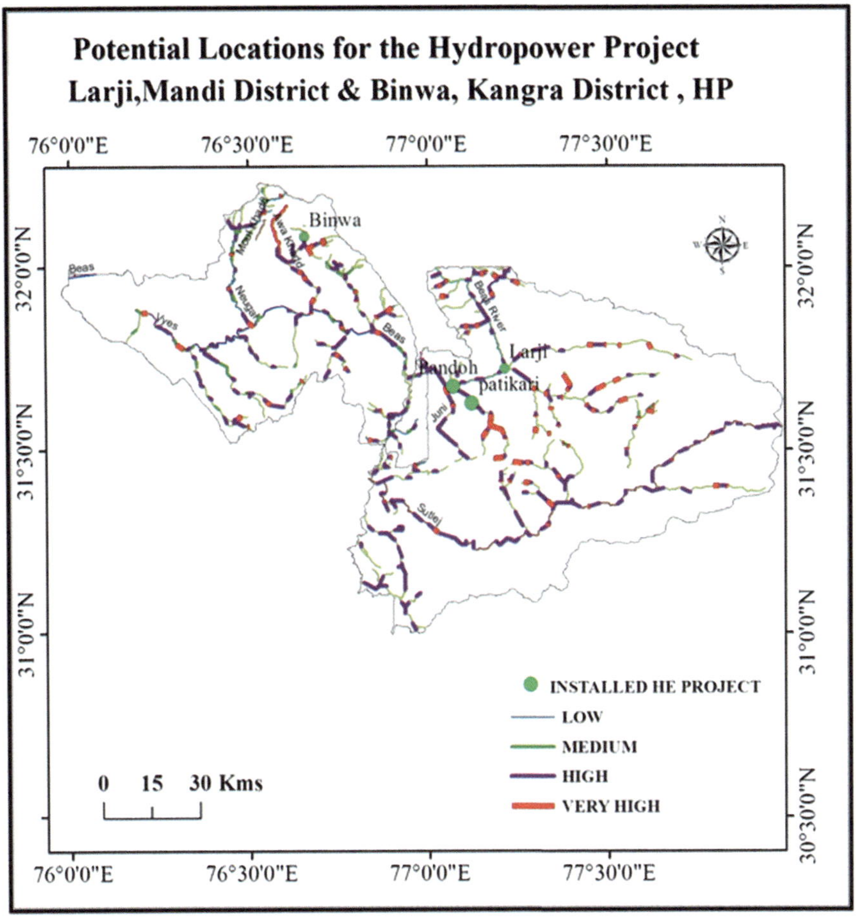

Fig. 7.8 Hydropower project potential zones

Acknowledgements The authors would like to express their gratitude to DOES & T, Himachal Pradesh, India for funding this study.

References

Acreman MC, Ferguson AJD (2010) Environmental flows and the european water system directive. Freshw Biol 55(1):32–48

Akter J (2010) Environmental flow assessment for the ganges river. Thesis submitted to the Department of Water Resources Engineering of Bangladesh University of Engineering and Technology, Dhaka

Al-Ruzouq R, Shanableh A, Yilmaz AG, Idris A, Mukherjee S, Khalil MA, Gibril MBA (2019) Dam site suitability mapping and analysis using an integrated GIS and machine learning approach. Water 11(9):1880
Chen J, Yang S, Li H, Zhang B, Lv J (2013) Research on geographical environment unit division based on the method of natural breaks (Jenks). Int Arch Photogramm Remote Sens Spat Inf Sci 3:47–50
De-sheng HU (2010) Water for the eco-environment: a perspective from international law. J Xi'An Jiaotong Univ 2:17
DHI (2006) Managed river flows for RHEP. DHI Water and Environment, New Delhi
Durbude DG, Jain CK, Singh O (2014) Assessment of E-flows for a river in southern india using hydrological index methods. J Indian Water Resour Soc 34(3):82–90
Earls J, Dixon B (2007) Spatial interpolation of rainfall data using ArcGIS: a comparative study, vol 31. In: Proceedings of the 27th Annual ESRI Int User Conference
EU: DIRECTIVE 2000/60/EC. Offic J Europ Comm
Glenn DM, Tabb A (2019) Evaluation of five methods to measure normalized difference vegetation index (NDVI) in apple and citrus. Int J Fruit Sci 19(2):191–210
Gnanachandrasamy G, Zhou Y, Bagyaraj M, Venkatramanan S, Ramkumar T, Wang S (2018) Remote sensing and GIS based groundwater potential zone mapping in ariyalur district, Tamil Nadu. J GeolSoc India 92(4):484−490
Goyal MK, Singh V, Meena AH (2015) Geospatial and hydrological modeling to assess hydropower potential zones and site location over rainfall dependent Inland catchment. Water Resour Manag 29(8):2875–2894
Halliday I, Robins J (2001) Environmental flows for sub-tropical estuaries. Final Report FRDC Project 2001(22)
Haque S, Kannaujiya S, Taloor AK, Keshri D, Bhunia RK, Ray PKC, Chauhan P (2020) Identification of groundwater resource zone in the active tectonic region of himalaya through earth observatory techniques. Groundw Sust Dev 10:P100337
Hassan I, Javed MA, Asif M, Luqman M, Ahmad SR, Ahmad A, Hussain B (2020) Weighted overlay based land suitability analysis of agriculture land in Azad Jammu and Kashmir using GIS and AHP. Pak J Agric Sci 57(6)
HMSO "Water Resources Act 1963" (1963)
Iftikhar S, Hassan Z, Shabbir R (2016) Site suitability analysis for small multipurpose dams using geospatial technologies. J Remote Sens GIS 1−13
Imran RM, Rehman A, Khan MM, Rahat M, Jamil UA, Mahmood RS, Ehsan RM (2019) Delineation of drainage network and estimation of total discharge using digital elevation model (DEM). Sci Technol 1(2):50–61
Jain SK, Kumar P (2014) Environmental flows in India: towards sustainable water management. Hydrol Sci J 59(3–4):751–769
Jasrotia AS, Kumar R, Taloor AK, Saraf AK (2019) Artificial recharge to groundwater using geospatial and groundwater modelling techniques in north western himalaya India. Arabian J Geosci 12(24):774
Jena S, Panda RK, Ramadas M, Mohanty BP, Pattanaik SK (2020) Delineation of groundwater storage and recharge potential zones using RS-GIS-AHP: application in arable land expansion. Remote Sens Appl Soc Environ 19:100354
Khan A, Govil H, Taloor AK, Kumar G (2020) Identification of artificial groundwater recharge sites in parts of yamuna river basin india based on remote sensing and geographical information system. Groundw Sustain Dev 100415
Kumar MG, Agarwal AK, Bali R (2008) Delineation of potential sites for water harvesting structures using remote sensing and GIS. J Indian Soc Remote Sens 36(4):323–334
Kumar P (2009) Environmental flow assessment for a hydropower project on a himalayan river. Thesis Ph.D., Indian Institute of Technology Roorkee, Roorkee
Lakshmi SV, Sarvani GR (2018) Selection of suitable sites for small hydropower plants using Geo-Spatial technology. Int J Pure App Math 119(17):217–240

Larentis DG, Collischonn W, Olivera F, Tucci CE (2010) Gis-based procedures for hydropower potential spotting. Energy 35(10):4237–4243
Nithya CN, Srinivas Y, Magesh NS, Kaliraj S (2019) Assessment of groundwater potential zones in chittar basin, southern india using GIS based AHP technique. Remote Sens Appl Soc Environ 15:100248
Otto JC, Prasicek G, Blöthe J, Schrott L (2018) GIS applications in geomorphology. In: Comprehensive Geographic Information Systems. Elsevier, pp 81 − 111
Paudel B, Zhang YL, Li SC, Liu LS, Wu X, Khanal NR (2016) Review of studies on land use and land cover change in Nepal. J Mount Sci 13(4):643–660
Saha A, Patil M, Karwariya S, Pingale SM, Azmi S, Goyal VC, Rathore DS (2018) Identification of potential sites for water harvesting structures using geospatial techniques and multi-criteria decision analysis. Int Arc Photogramm Remote Sen Spat Inf Sci
Sahu I, Prasad AD (2018) Assessment of hydro potential using integrated tool in QGIS ISPRS annals photogramm, remote sens spatial. Inf Sci 45:115–119
Samanta S, Aiau SS (2015) Spatial analysis of renewable energy in papua new guinea through remote sensing and GIS. Int J Geosci 6(08):853
Sammartano V, Liuzzo L, Freni G (2019) Identification of potential locations for Run-of-River hydropower plants using a GIS-based procedure. Energies 12(18):3446
Sarkar T, Kannaujiya S, Taloor AK, Ray PK, Chauhan P (2020) Integrated study of GRACE data derived interannual groundwater storage variability over water stressed indian regions. Groundw Sustain Dev P100376. https://doi.org/10.1016/j.gsd.2020.100376
Singh AK, Jasrotia AS, Taloor AK, Kotlia BS, Kumar V, Roy S, Ray PK, Singh KK, Singh AK, Sharma AK (2017) Estimation of quantitative measures of total water storage variation from GRACE and GLDAS-NOAH satellites using geospatial technology. Quat Int 444:191 − 200
Soni V, Shekhar S, Singh D (2013) Environmental flow for monsoon rivers in india: the yamuna river as a case study [online]. Submitted on 12 June 2013 to the Cornell University. http://arxiv.org/ftp/arxiv/papers/1306/1306.2709.pdf. Accessed 4 Nov 2013
Souchon Y, Keith P (2001) Freshwater fish habitat: science, management, and conservation in france. Aquat Ecosyst Health Manage 4(4):401–412
Sundaray P (2013) Environmental flow (Doctoral dissertation)
Tarife RP, Tahud AP, Gulben EJG, Macalisang HARCP, Ignacio MTT (2017) Application of geographic information system (GIS) in hydropower resource assessment: a case study in misamis occidental. Int J Env Sci Develop 8(7):507
Van Wyk E, Breen CM, Roux DJ, Rogers KH, Sherwill T, van Wilgen BW (2006) The ecological reserve: towards a common understanding for river management in south africa. Water SA 32(3):403–409
Wang Q, MIkiugu MM, Kinoshita I (2014) A GIS-based approach in support of spatial planning for renewable energy: a case study of Fukushima Japan. Sustainability 6(4):2087–2117
WWF (Worldwide Fund for Nature) (2012) Summary report. Assessment of environment flows for upper ganga basin. WWF-India, New Delhi
WWF (Worldwide Fund for Nature) (2013) Environmental flows for Kumbha 2013. Triveni Sangham. WWF-India, Allahabad
Xu H (2006) Modification of normalised difference water index (NDWI) to enhance open water features in remotely sensed imagery. Int J Remote Sen 27(14):3025–3033

Chapter 8
Reassessing the Karakoram Through Historical Archives

Rakesh Bhambri, Pritam Chand, Marcus Nüsser, Prashant Kawishwar, Amit Kumar, Anil K. Gupta, Akshaya Verma, and Sameer K. Tiwari

Abstract The Karakoram region is an integral part of the Hindu Kush–Karakoram–Himalaya (HKH) mountain system and forms the great divide between the drainage of the Indus and Tarim basins. This region has a substantial geographical, historical, economical, cultural, and geopolitical importance owing to its location, rugged topography, distinct climate regime, and the presence of numerous glaciers. It acts as a water tower where its unique assemblage paved the way for developing ancient trade routes facilitating cultural exchange between Central Asia and the Indian subcontinent. It has attracted a large number of expeditions and scientific studies for more than 200 years, bringing significant advances in geographical knowledge. However, the spatial extent of the Karakoram in many studies particularly during the 2010s exhibits a large inconsistency. We, therefore, complied ≥100 maps and vast literature to present the evolution of the geographical term 'the Karakoram' and its spatial extent based on historical archives. We also provide a digital outline of the Karakoram region using GIS tools based on descriptive enumerations of the Royal Geographical Society (RGS) and the Survey of India (SoI) which we hope will provide new insight

Electronic supplementary material The online version of this chapter (https://doi.org/10.1007/978-3-030-47660-1_8) contains supplementary material, which is available to authorized users.

R. Bhambri · M. Nüsser
Department of Geography, South Asia Institute, Heidelberg University, Voßstraße 2/4130, 69115 Heidelberg, Germany

P. Chand (✉)
Department of Geography, School of Environment and Earth Sciences, Central University of Punjab, VPO-Ghudda, Bathinda 151401, India
e-mail: pritam.chand@cup.edu.in

P. Kawishwar
Chhattisgarh Council of Science and Technology, Vigyan Bhavan, Vidhansabha Road, Raipur 492014, India

R. Bhambri · A. Kumar · A. Verma · S. K. Tiwari
Wadia Institute of Himalayan Geology, 33 GMS Road, Dehradun 248001, India

A. K. Gupta
Indian Institute of Technology, Kharagpur 721302, India

A. Saikia and P. Thapa (eds.), *Environmental Change in South Asia*,
https://doi.org/10.1007/978-3-030-47660-1_8

for a wide range of scientific disciplines, including geography, glaciology, geology, history, and cartography.

Keywords Karakoram · Survey of India · Royal Geographical Society · Historical geography · GIS

8.1 Introduction

The Karakoram region is an important part of the Hindu Kush–Karakoram–Himalaya (HKH) mountain system. This region lies between 34° 07′–36° 51′ N latitudes and 73° 43′–78° 20′ E longitudes. The Karakoram forms the great divide between the drainages of Central Asia (e.g., Tarim basin) and the Arabian Sea (Indus basin) over a length of approximately 700 km. Several of the largest glaciers of the world outside the polar regions are located in this area (e.g., Siachen, Biafo, Baltoro, Batura, Hispar). These glaciers significantly contribute to the runoff from the Karakoram region (Reggiani et al. 2017) and cater to the vast population living in this region and its downstream areas. The network of ancient trade routes (e.g., Leh–Yarkand–Kashgar) between Central Asia and the Indian subcontinent also pass across the Karakoram (Hayward 1870; Middleton 1984; Kreutzmann 2007). The Karakoram region has recently regained the attention of the scientific community due to its nearly balanced mass budgets (Hewitt 2005; Cogley 2012; Gardelle et al. 2012, 2013; Kääb et al. 2012, 2015; Bolch et al. 2017), increased glacier surge events (Barrand and Murray 2006; Bhambri et al. 2013, 2017; Rankl et al. 2014; Paul 2015; Quincey et al. 2015), decrease of water discharge in the recent past (Archer 2003; Reggiani et al. 2017), unique seasonal cycle of snowfall that is dominated by non-monsoonal winter precipitation (Kapnick et al. 2014), and reported cooling summertime surface air temperatures (i.e., linked with the Karakoram Vortex) (Forsythe et al. 2017; Li et al. 2018) over this region. Some studies have presented the extent of the entire Karakoram or its parts on maps (e.g., Cogley 2011; Bolch et al. 2012; Gardelle et al. 2012, 2013; Bhambri et al. 2013; Rankl et al. 2014; Pratap et al. 2016; RGI Consortium 2017). However, delineations of the Karakoram in most of these studies are not congruent (Fig. 8.1). Most studies include surrounding areas like the Aghil, Chang Chenmo, Nanga Parbat, and Ladakh mountains on their maps and spatial calculations (e.g. Cogley 2011; Bolch et al. 2012; Gardelle et al. 2012, 2013; Bhambri et al. 2013; Rankl et al. 2014; Pratap et al. 2016; RGI Consortium 2017; Zhou et al. 2017). This suggests that there is no international standardization in the boundary of the Karakoram. A consistency in the extent of the Karakoram region is thus required.

(1) to quantify, analyze and compare databases of natural and cultural resources for scientific investigation on a common platform and;
(2) to understand the spatial distribution and temporal changes of resources in the region;
(3) for harmonization of scientific studies.

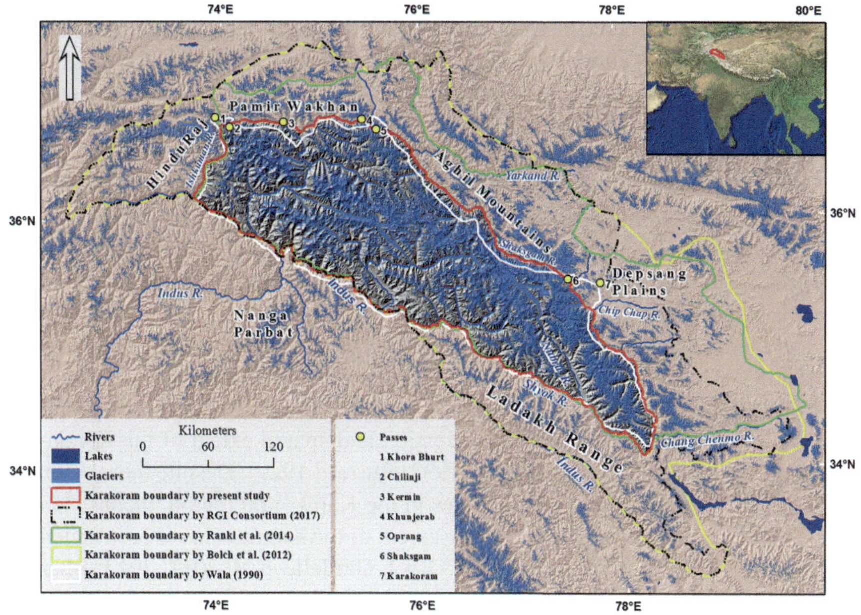

Fig. 8.1 Overview of the Karakoram and surrounding area (Hindu Raj, Pamir Wakhan, Aghil mountains, Depsang Plains, Ladakh, and Nanga Parbat Himalaya). The Karakoram boundary has been prepared using the description given by the Royal Geographical Society and Survey of India nomenclature (Mason 1938). The hill-shaded background in this figure was generated by using the Shuttle Radar Topography Mission (SRTM) dataset (90 × 90 m) provided by the CGIAR Consortium for Spatial Information (http://srtm.csi.cgiar.org/). The glacier outlines were acquired from the Randolph Glacier Inventory version 6.0 (RGI Consortium 2017). The digital boundaries of Karakoram have been acquired from Bolch et al. (2012) and RGI Consortium (2017) whereas the outline of Karakoram was digitized from Wala (1990) and Rankl et al. (2014)

The well-defined spatial extent of the Karakoram region is, therefore, a subject of considerable multi-disciplinary interest in geography, geology, glaciology, environmental studies, history, and cartography. A standardized map is important to represent the spatial extent of the high mountain region together with further information including locations, place names, and a two-dimensional delineation. Some studies believed that the local name has always existed and mapmakers only have to attach it to a feature or place on their maps (e.g., Bhattacharji 2010). Moreover, the occurrence, positioning, and meanings of toponyms (geographical names) on maps are at the same time indicators of relations between spatial identities. Thus, cartographical representations of the territories are always a part of political power constellations and administrative functionality. In this context, the Karakoram region presents a textbook example for the continuity of contested territoriality from colonial rule to the present day. The politically sensitive territoriality resulting from conflicts between the nation-states of India, Pakistan, and China can be regarded as a key factor for the difficulty of mapping the Karakoram spatial configuration and boundaries (Baghel and Nüsser 2015). Recent progress in Geographical Information System (GIS) and free access to high-resolution satellite images (e.g., Google Earth) has led to a marked improvement in map-making and generated a large digital database that facilitates spatial and non-spatial data analyses.

The Karakoram is a Turkic term, which means black gravel (Burrard 1929; Mason 1955). This term possibly was first used by traders who travelled along the historical trade route between India and Yarkand (Mason 1928). This is an esoteric name for one of the world's highest and most heavily glacierized mountain ranges. It is also assumed that the name refers to the Karakoram Pass with its dark mountain characteristics located approximately 180 km from the Biafo Glacier in snow clad mountains between the Aksai Chin and the Depsang Plains (Dyhrenfurth 1955). An extensive discussion on Karakoram nomenclature has been reported in the literature during the 1920s and 1930s. Leading geographers, explorers, and surveyors to propose a solution to problems of regional classification and spatial extent of the Karakoram mountain belt (Mason 1928, 1930a, b, 1938; Burrard 1929). Despite these efforts to delineate and categorize mountain ranges in the Karakoram region, inconsistencies can be discerned in the delineation of the extent of the Karakoram in recent scientific analyses (e.g. Cogley 2011; Bolch et al. 2012; Gardelle et al. 2012, 2013; Bhambri et al. 2013; Rankl et al. 2014; Pratap et al. 2016; RGI Consortium 2017; Zhou et al. 2017). Against this background, this article aims to reassess the spatial extent of the Karakoram (Fig. 8.1). Some notable studies have reported the historical progress of exploration and mapping of the Karakoram (Hedin 1922; Dyhrenfurth 1955; Mason 1955; Desio 1991; Heichel 2010), whereas only a few studies used the outline of the Karakoram based on the definition given by the Royal Geographical Society (RGS) and the Survey of India (SoI) (e.g., Bhambri et al. 2017; Groos et al. 2017). However, no attempt has been made to provide the outline of the Karakoram in digital GIS format in a public domain based on the definition given by RGS and SoI during the 1938 Karakoram conference.

Our study, therefore, reviews, analyses, and synthesizes the historical development of the term "Karakoram", its spatial extent, and the contribution of various organizations in the progress of mapping the Karakoram. We also highlight the gap areas in the delineations of the Karakoram and present a widely accepted outline of the Karakoram based on descriptive enumerations of RGS and SoI that may help various scientific disciplines especially in the context of quantitative analyses.

8.2 Data and Methodology

The main approach in this study is to collate and synthesize materials from historical archives including published reports, scientific articles, monographs, and maps since the fourteenth century and to analyze excerpts from these sources in order to trace changes in toponyms and spatial extent of the Karakoram. We acquired digital boundaries of the Karakoram generated by Bolch et al. (2012), RGI Consortium (2017), and digitized Karakoram outlines from the maps created by Wala (1990) and Rankl et al. (2014). We also used GIS tools to map the digital outline of the Karakoram region based on the description given by the RGS and SoI during the 1938 Karakoram conference (Mason 1938). Digital datasets such as Shuttle Radar Topography Mission (SRTM v4) digital elevation model (DEM) and Google Earth high-resolution images were used to generate the outline of the Karakoram.

During the 1938 Karakoram conference, latitude, longitude, river extents, and passes were used to define the descriptive boundary of the Karakoram (Mason 1938). Rivers like Indus, Gilgit, Ishkoman, Chapursan, Kilik, Khunjerab, Oprang, Shaksgam, Yarkand, Rimo, and Chip-chap were treated as boundaries of the Karakoram Range (Fig. 8.1). In addition, the location of passes (i.e., Chilinji, Kermin, Khunjerab, Oprang, and Shaksgam) proposed by the Karakoram conference in 1938 (Mason 1938) was acquired from high-resolution Google Earth images (Table 8.1). More details on the Karakoram boundary established and described by RGS and SoI during the 1938 Karakoram conference are presented in Sect. 8.3. We followed an automatic procedure using GIS to generate the drainage networks from SRTM DEM for delineating the rivers (Khan et al. 2014). The steps include mainly artefact fills in SRTM DEM and flow path analysis using ArcGIS 10.5 software. DEMs have frequent topographic depressions and possibly even flat areas. These artefacts can generate errors for flow-path routing in the DEM, as these features have no downslope neighbours. Therefore, we used the fill tool using ArcGIS 10.5 for removing artefacts from SRTM DEM and the flow direction tool for river flow path analysis. River boundaries were joined manually with passes. This boundary was compared with a previously defined spatial extent by other studies (e.g., Bolch et al. 2012; RGI Consortium 2017).

8.3 Revisiting Mapping and Extent of the Karakoram

8.3.1 Progress of the Karakoram Exploration and Mapping Before the 1800s

The earliest references to the mapping of Himalayan mountain ranges can be traced back since the fifteenth century, whereas mapping of surrounding mountain regions of the Karakoram probably got underway during the late sixteenth century. Many ancient maps such as South Asia by Ptolemy from the year 1410 and maps of India by Giacomo Gastaldi from the year 1561 did not show the location of the Karakoram and nearby places (Table 8.1). Most earlier maps prepared during the seventeenth and eighteenth centuries included adjacent populated mountain regions (e.g., Baltistan and Little Tibet) instead of the Karakoram. The mountain region of the Karakoram was possibly for the first time, presented on a map of northern India by Antonio de Montserrat, produced in the year 1582 (Hedin 1922). The Masarvor (presently known as Manasarovar lake) and Casmir (present day Kashmir) are also located on this map (Supplementary Fig. 8.1). The mountains to the west and south are presented as

Table 8.1 Chronology of mapping of Karakoram region

Explorer/cartographer	Country	Year of map publication/exploration	Map title/area	Scale	References
Ptolemy's	Greece	1410	South Asia	No scale	Phillimore (1950)
Gastaldi	Italy	1561	Central Asia	No scale	Phillimore (1945)
Antonio de Montserrat	Portugal	1582	India	No scale	Hedin (1922), Phillimore (1945)
Mercator	Unknown	1612	India	Presented in other language	Phillimore (1945)
Bernier	France	1656–1668	Map of Mogul Empire	–	www.raremaps.com
Visscher	Netherlands	1670	India Orientalis (East India)	Presented in other language	www.raremaps.com
Chatelain	France	1719	East Asia	Presented in other language	www.bartelegallery.com
Rennell	United Kingdom	1792	Countries between the Ganges and Caspian Sea	1.5 inches to a degree	Rennell (1793), Phillimore (1945)
Rennell	United Kingdom	1792	Countries between Delhi and Candahar	1.5 inches to a degree	Rennell (1793), Phillimore (1945)
Arrowsmith's	United Kingdom	1801	Asia map	No scale	Hedin (1922), Phillimore (1945)
Macartney	United Kingdom	1815	Map of Kingdom of Caubul	50 miles to 1 inch	Elphinstone (1815)
Moorcroft and Trebeck	United Kingdom	1841	Map of the Himalayan Provinces of Hindustan, The Punjab, Ladakh, Kashmir, Kabul, Kunduz and Bokhara	No scale	Moorcroft and Trebeck (1841)

(continued)

Table 8.1 (continued)

Explorer/cartographer	Country	Year of map publication/exploration	Map title/area	Scale	References
Strachey	United Kingdom	1851	Map of Ladak East with parts of Balti and Monyul	8 miles to 1 inch	–
Strachey	United Kingdom	1854	Map of west Nari	50 miles to 1 inch	Strachey (1853)
Atkinson	Unknown	1860	Central Asia	437 miles to 1 inch	Atkinson (1860)
Godwin-Austen	United Kingdom	1864	Sketch Map of the glaciers of the Mustakh range (Trans-Indus) and valley of Skardo	10 Miles to 1 Inch	Godwin-Austen (1864)
Johnson	United Kingdom	1865	Map of the Country between Leh and Ilchi	34 Miles to 1 Inch	Johnson (1867)
Robert Shaw	United Kingdom	1871	Map of the route from Leh to Yarkand and Kashghar	56 Miles to 1 Inch	Shaw (1871)
Hayward	United Kingdom	1871	Sketch Map of the Trans-Indus Countries including Gilgit, Dilail, Yassin	22 Miles to 1 Inch	Hayward (1870)
Survey of India	India	1874	Route of Kishen Singh in Western Tibet	32 Miles to 1 Inch	http://pahar.in/
Montgomerie	United Kingdom	1874	Route Map for Western Himalayas and Northern India	32 Miles to 1 Inch	http://pahar.in/
Drew	United Kingdom	1874	Territories of Maharaja of Jummoo and Kashmir	16 Miles to 1 Inch	Drew (1875)

(continued)

Table 8.1 (continued)

Explorer/cartographer	Country	Year of map publication/exploration	Map title/area	Scale	References
Survey of India	India	1874	Route Map for the Western Himalayas, Kashmir, Punjab and Northern India, W. Portions of Afghanistan, Baluchistan and C. '(including the Former Kashmir Route Map)'	32 Miles to 1 Inch	–
Drew	United Kingdom	1875	Jummoo and Kashmir Snow Map	32 Miles to 1 Inch	Drew (1875)
Marsh	United Kingdom	1876	Sketch Map of countries surrounding Gilgit	32 Miles to 1 Inch	Marsh (1876)
Survey of India	India	1876	Explorations in connection with GTS by Mullah (Trans-Himalayan Exploration Map 11)	12 Miles to 1 Inch	http://pahar.in/
Survey of India	India	1884	Dardistan or Shinaki including Chilas by Ahmad Ali of SoI	4 Miles to 1 Inch	http://pahar.in/
Survey of India	India	1888	Chitral Hunza and parts of Wakhan and Kafiristan by Woodthorpe	12 Miles to 1 Inch	http://pahar.in/
Younghusband	United Kingdom	1890	Map to illustrate the exploitations of Captain Younghusband, King's Dragoon Guards on the Northern Frontier of Kashmir	–	http://pahar.in/
Knight	United Kingdom	1891	Portion of Kashmir and Adjacent States	31 Miles to 1 Inch	Knight (1893)
Conway	United Kingdom	1892	The Hispar Glacier	1:126,720	Conway (1894)
Conway	United Kingdom	1892	Biafo and Baltoro Glaciers	1:126,720	Conway (1894)

(continued)

Table 8.1 (continued)

Explorer/cartographer	Country	Year of map publication/exploration	Map title/area	Scale	References
Leitner	United Kingdom	1893	Map of Dardistan and the Pamirs	1:4,000,000	http://pahar.in/
Younghusband	United Kingdom	1895	Chitral and Adjacent Countries	23.6 Miles to 1 Inch	Younghusband (1895)
Workman and Workman	USA	1899	Map of Biafo Glacier	4 Miles to 1 Inch	Workman and Workman (1900)
Workman	USA	1903	Part of the Karakoram Himalaya, including the Chogo Lungma, Alchori, Hoh Lumba and Sosbon glaciers. Surveyed by the Bullock Workman Expedition, 1903	1:250,000	Workman (1906)
Workman and Workman	USA	1906	Nun-Kun Massif and surrounding region in the Kashmir Himalaya	1:175,000	Workman and Workman (1908)
Workman	USA	1908	Hispar Glacier and tributaries in the Karakoram Range	1:150,000	Workman (1910)
Longstaff	United Kingdom	1910	Exploration in the Eastern Karakoram	1:500,000	Longstaff (1910)
Duke of Abruzzi	Italy	1910	Portion of Karakoram Range by Abruzzi expedition	1:100,000	http://pahar.in/
Neve	United Kingdom	1910	Diagram of the Ranges of North Kashmir	1:100,000	Neve (1910)
Duke of Abruzzi	Italy	1911	Baltoro and Godwin-Austen Glacier	1:125,000	De Filippi (1911)
Workman and Workman	USA	1911–1912	Map to illustrate the Hushe and Kondus glacier system	1:200,000	Workman (1912)

(continued)

Table 8.1 (continued)

Explorer/cartographer	Country	Year of map publication/exploration	Map title/area	Scale	References
Workman	USA	1912	The Siachen or Rose Glacier and tributaries (Eastern Karakoram)	1:175,000	Workman (1914)
De Filippi	Italy	1913–1914	Spedizionede Filippi II ghiaccialorimu e la zona d'AltipianoFino al passo Caracorum	1:100,000	De Filippi (1924)
De Filippi	Italy	1913–1914	Depsang Plains by De Filippi Expedition	1:250,000	De Filippi (1932)
De Filippi	Italy	1914	Upper Yarkand Valley by De Filippi Expedition	1:250,000	De Filippi (1915)
Survey of India	India	1916	Sheet No 52 Leh	1/1000000	http://pahar.in/
Survey of India	India	1916	Sheet No 43 Srinagar	1/1000000	http://pahar.in/
Survey of India	India	1918	Sheet No 42 Pamirs	1/1000000	http://pahar.in/
Visser	Netherlands	1926	Explorations in the Karakoram	1:500,000	Visser (1926)
Younghusband	United Kingdom	1926	Problem of Shaksgam	1:1,000,000	Younghusband (1926)
Survey of India	India	1926	Sheet No 51, Yarkand	1/1000000	http://pahar.in/
Mason	United Kingdom	1927	Shaksgam Valley and Aghil Range	1:250,000	Mason (1927a)
Mason	United Kingdom	1927	Stereo Survey of Shaksgam	1:100,000	Mason (1927b)
Morris	United Kingdom	1927	Map of valleys and glaciers in Hunza	1:400,000	Morris (1928)
Survey of India	India	1928	Shaksgam and Upper Yarkand Valleys	1/253440	Mason (1928)
Survey of India	India	1929	The range lines and names proposed by Mason in his Shaksgam report	No scale	Anonymous (1929)

(continued)

Table 8.1 (continued)

Explorer/cartographer	Country	Year of map publication/exploration	Map title/area	Scale	References
Survey of India	India	1930	Diagram of proposed range-names in the Karakoram	No scale	Mason (1930a)
Gunn	United Kingdom	1930	Upper Shyok Glaciers and Gapshan Lake	1/253440	Gunn (1930)
Dainelli	Italy	1932	Glaciers of the Eastern Karakoram	1:750,000	Dainelli (1932a,b)
Visser and Visser	Netherlands	1933	Map of the Visser-Expeditions (Explored areas in Karakoram and Aghil mountains)	1/200000	Visser and Visser (1938)
Visser	Netherlands	1934	Karakoram and Turkistan	1:500,000	Visser (1934)
Survey of India	India	1934	Karakoram-Diagram of Peaks and Ranges	No scale	http://pahar.in/
Schomberg	United Kingdom	1936	Shingshal Mustagh area	1/253440	Schomberg (1936)
Spender and colleagues	United Kingdom	1937	Parts of great Karakoram and Aghil mountains	1:250,000	Shipton et al. (1938)
Mason	United Kingdom	1938	Sketch to illustrate the Karakoram	12 Miles to 1 Inch	Mason (1938)
Shipton	United Kingdom	1939	Hispar-Biafo Glacial System	1/253440	Mott (1950)
Royal Geographical Society	United Kingdom	1939	The Karakoram	1:750,000	http://pahar.in/
Dyhrenfurth	Germany and Switzerland	1939	Baltoro, Ein Himalaya-Buch	1:175,000	Wala (1990)
Lombardi and team	Italy	1954	Stak and Turmik Valleys by Italian Karakorum Expedition	1:100,000	Desio (1991)
Kurz	Switzerland	1954	Karakoram	1:750,000	Wala (1990)

(continued)

Table 8.1 (continued)

Explorer/cartographer	Country	Year of map publication/exploration	Map title/area	Scale	References
Kick	Germany	1954	Chogo-Lungma glacier in Baltistan	1:100,000	Kick (1956)
Italian Military Geographic Institute	Italy	1955	K2 by Italian Expedition to Karakorum Himalaya	1:12,500	Desio (1991)
Lombardi and team	Italy	1957	Stak Valley of Karakorum Himalaya	1:50,000	Desio (1991)
USA Army map service	USA	1960s	India and Pakistan series U 502, sheet NI 43–3, NI 43–4, NI 43–2, NI 43–8, NI 43–13, NJ 43–14 prepared by Army map service	1:250,000	http://pahar.in/
Schneider	Germany	1967	Minapin (Rakaposhi range)	1:50,000	Wala (1990)
Finsterwalder	Germany	1969	Hunza-Karakorum 1:100,000: Topographische Ergebnisse der Deutsch-Österreichischen Himalaya-Karakorum-Expedition 1954 und der Deutschen Karakorum-Expedition 1959.	1: 100 000	Finsterwalder (1996)
Hewitt	Canada	1969	Distribution of surge glaciers in the Karakoram	No scale	Hewitt (1969)
Hartmann	Switzerland	1970	Erinnerungen an die Exkursionen am Byafo Gyang (Karakoram)	?	Wala (1990)
Wala	Poland	1974	Topography of the Hispar Muztagh	?	Wala (1990)
Wala	Poland	1974	Hispar Muztagh, Orographical sketch map	1:125,000	Wala (1990)
Wala	Poland	1976	Orographical sketch map of the Karakoram, 3rd edition	1:250,000	http://pahar.in/

(continued)

Table 8.1 (continued)

Explorer/cartographer	Country	Year of map publication/exploration	Map title/area	Scale	References
Kobe university expedition	Japan	1976	Sherpi Kangri	1:150,000	Wala (1990)
Miyamori	Japan	1976 and 1977	Orographical maps series sheet No.1,No.2,No.3,No.4	1:200,000	Wala (1990)
Kurz	Switzerland	1977	Karakoram	1:750,000	Wala (1990)
Rampini	Italy	1977	Karakoram sistema latok ghiacciaio Baintha	1:20,000	Wala (1990)
Institute of glaciology, cryopedology and desert research, academia scinialanchow, china	China	1978	Batura Glacier map	1:60,000	Institute of glaciology (1978)
Survey of Pakistan	Pakistan	1983	Survey of Pakistan Hunza map, sheet Nj-43/SW	1:500,000	Wala (1990)
Wala	Poland	1984	Batura Muztagh	1:125,000	Wala (1990)
Survey of Pakistan	Pakistan	1986	Pakistan Khunjerab pass, sheet NJ-43/SE	1:50,000	Wala (1990)
Wala	Poland	1987	Orographical sketch map of Batura wall-pasursar-shisparesar	1:100,000	Wala (1990)
Miyamori	Japan	1988	The Baltoro Glacier, inserted folded map for the mountaineering	?	Wala (1990)
Wala	Poland	1988	Karakoram-Batura Wall, Pasu Sar, Shispare Sar	1:100,000	http://pahar.in/
Wala	Poland	1990	Orographical sketch map of the Karakoram (two sheets)	1:250,000	http://pahar.in/

(continued)

Table 8.1 (continued)

Explorer/cartographer	Country	Year of map publication/exploration	Map title/area	Scale	References
Baume	Unknown	1990	Trekking Maps of the Karakoram (four sheets)	1:200,000	Baume (1990)
Wala	Poland	1990	Haramosh Karakoram, Central Karakoram	1:100,000	http://pahar.in/
Wala	Poland	1994	8000 m Peaks of Karakoram	1:50,000	http://pahar.in/
Wala	Poland	2005	Hispar Muztagh, Central Karakoram	1:125,000	http://pahar.in/
Wala	Poland	2005	Rakaposhi-Malubiting Mountains	8 Miles to 1 Inch	http://pahar.in/
Wala	Poland	2006	Haramosh Karakoram, Central Karakoram	1:125,000	http://pahar.in/
Wala	Poland	2006	Karakoram-Spantik, Sosbun Mountains	1:125,000	http://pahar.in/
Wala	Poland	2007	Karakoram- Gulmit Tower	200 m to 1 cm	http://pahar.in/
Wala	Poland	2007	Western Karakoram-Batura Muztagh	480 m to 1 cm	http://pahar.in/
Iturrizaga	Germany	2011	Overview of Karakoram and glacier investigated in study	25 km to 1 cm	Iturrizaga (2011)
Hewitt and Liu	Canada and China	2011	Location map of the Upper Indus and Yarkand basins showing the distribution of glaciers associated with large ice dams and known for large outburst floods	12 km to 1 cm	Hewitt and Liu (2010)
Wala	Poland	2012	Karakoram Latok and Choktoi Mountain Subgroups	2000 m to 1 cm	http://pahar.in/

(continued)

Table 8.1 (continued)

Explorer/cartographer	Country	Year of map publication/exploration	Map title/area	Scale	References
Wala	Poland	2012	Karakoram- Masherbrum Khane Valley	560 m to 1 cm	http://pahar.in/
Wala	Poland	2013	Karakoram- Ghujerab Ghidims Valley by Jerzy Wala	700 m to 1 cm	http://pahar.in/
Wala	Poland	2013	Karakoram- Wirjerab Muztagh by Jerzy Wala	2200 m to 1 cm	http://pahar.in/
Wala and Majer	Poland	2014	Karakoram-Shuijerab Mountain Group	2300 m to 1 cm	http://pahar.in/
Wala and Majer	Poland	2014	Karakoram-Shuijerab, Subgroup M2 by Jerzy Wala, Janusz Majer	1300 m to 1 cm	http://pahar.in/
Wala and Majer	Poland	2014	Karakoram-Shuijerab, Subgroup M3 by Jerzy Wala, Janusz Majer	1300 m to 1 cm	http://pahar.in/
Wala and Majer	Poland	2014	Karakoram-Shuijerab, Subgroup M4 by Jerzy Wala, Janusz Majer	1300 m to 1 cm	http://pahar.in/
Wala and Majer	Poland	2014	Karakoram-Shuijerab, Subgroup M5 by Jerzy Wala, Janusz Majer	1300 m to 1 cm	http://pahar.in/
Wala and Majer	Poland	2014	Karakoram-Shuijerab, Subgroup M6 by Jerzy Wala, Janusz Majer	1300 m to 1 cm	http://pahar.in/
Gibson and team	United Kingdom	2016	Changes in glacier surface cover on Baltoro Glacier, Karakoram, North Pakistan, 2001–2012	1800 m to 1 cm	Gibson et al. (2017)

Fig. 8.2 **a** Map of the Mogul Empire made by Francois Bernier during travels in the Mogul Empire (1656–1668 A.D.); **b** part of Indiae Orientalis map prepared by Nicolaes Visscher during ~1670 A.D.; **c** Map of East Asia by Chatelain (1719 A.D.); Map of countries between Delhi and Candahar by Rennell (1793 A.D.)

IMAYS (presently known as Himalaya). A map of the Mogul Empire made by Francois Bernier (1656–1668) showed the position of Tibet Minor (Baltistan), Eskerdou (presently known as Skardu), and Kachger (presently known as Kashgar). Nicolaes Visscher also located all these places of interest on the map of Indiae Orientalis (East India) from the year 1670 (Fig. 8.2). However, the problem of limited positional accuracy in these early maps has been redressed by the availability of recent high-resolution satellite imagery sources (e.g., Google Earth and Google maps). The collated maps (Table 8.1) suggest that the term Karakoram did not appear on maps that were generated before the nineteenth century.

8.3.2 Progress of the Karakoram Exploration and Mapping from 1800 to 1900

Cartographic representation of the Karakoram toponyms and its spatial extent started during the eighteenth century. The term Mus Tag (Karakoram Range) was used for the very first time by Arrowsmith on his Asia map (Fig. 8.3) in 1801 (Hedin 1922). This

Fig. 8.3 Map of Asia by Arrowsmith in 1801 (Hedin 1922)

map became the base reference for several expeditions and maps of the Himalaya–Karakoram Range during the first half of the nineteenth century and subsequently improved in terms of information content and accuracy of spatial information (e.g., Moorcroft and Trebeck 1841). However, the term Mus Tag or Muztagh is not regarded as a name but a noun in Balti-vocabulary, used in Baltistan, denoting "snow or ice-mountain" (Burrard 1929). Mir Izzet Ullah (an agent of William Moorcraft) two centuries ago, began the exploration of the Karakoram region from Kashmir to Kashgar in 1812 via the Karakoram Pass (Mason 1955). He presented a narrative account of his 'Karakoram' journey in Persian and this was translated in the Calcutta Oriental Quarterly Magazine in 1825 and reproduced in the Journal of the Royal Asiatic Society in 1843 (Ullah 1843). The Karakoram term was thus established in the literature and also located on maps during the first half of the nineteenth century.

Several explorers prepared regional route maps for parts of the Karakoram region (e.g., Strachey 1853; Hayward 1870) while others generated maps in the context of the Great Trigonometrical Survey (GTS e.g., Godwin-Austen 1864) during the second half of the nineteenth century. The three German brothers Adolph, Hermann, and Robert Schlagintweit investigated numerous valleys in the Karakoram (e.g., Shigar, Braldo, Hushe, Kundos, Nubra, and Shyok) and adjacent mountain regions in the 1850s. They produced several regional maps and landscape panoramas for parts of the Karakoram (Kick 1960; Brescius et al. 2015). A clear and concise account of the glaciers (Baltoro, Biafo, Chogo Lungma) in the Karakoram (Mustagh Range) was published for the first time by Godwin Austen in 1864 during the GTS (Godwin-Austen 1864). This black and white plane-table survey map showed relief of glacier terrain by a system of hachure (Supplementary Fig. 8.2).

Godwin-Austen, a subordinate of James Thomas Walker, Superintendent of the GTS, SoI used the term 'Mustagh Range' for his glacier survey in the great Karakoram Range during the 1860s (Godwin-Austen 1864). However, Montgomerie who was also a subordinate of Walker used the letter 'K' for defining the peaks of the Karakoram. The use of the letter K is well known in the history of discovery with the second highest peak in the world, which still bears the initial identification symbol K2 as an internationally recognized name. This shows that both terms, Karakoram and Muztagh, were adopted by the SoI during the 1850–1860s. SoI discontinued the use of the term 'Muztagh' in the 1870s and was hesitant to adopt the Karakoram term (Burrard 1929). In 1876, Walker, the Surveyor General of India, approved the name Karakoram to generate Montgomerie's charts of triangulation. Since then the maps of the Government of India have represented the great range as the Karakoram. A correct regional terminology of the Karakoram peaks was also discussed during the last decade of the nineteenth century (Godwin-Austen and Conway 1893a, b). However, it is almost certain that no map presented the entire Karakoram with its outline during the nineteenth century.

8.3.3 Progress of the Karakoram Exploration and Mapping from 1900 Onwards

Several explorers have conducted numerous individual surveys of glacier areas during the first half of the twentieth century that contributed significantly to the mapping of the Karakoram (Table 8.1). The herring-bone arrangement of the transverse valleys on either side of the Karakoram range by the main rivers including the Indus, Shyok, and Yarkand and its tributaries were first attributed to regional geology by Neve (1910) (Supplementary Fig. 8.3). Younghusband (1926) identified the problem of unsurveyed areas of the Shaksgam Valley based on his first small-scale map (1:100,000) of the Karakoram and compilation of previous surveys and explorations (Table 8.1). Since then several exploration and mountaineering expeditions (e.g., Shipton 1938) remarkably filled the unexplored gaps of the mountain region (Table 8.1). Baghel and Nüsser (2015) provided a historical exploration of the Siachen Glacier and presented a number of toponyms for the region.

A long discussion on the spelling of Karakoram also took place in the 1930s. Burrard changed the spelling of the name Karakoram to the Karakorum without providing any reason (Burrard and Reynolds 1935), whereas he adopted the term Karakoram in his previous work (e.g., Burrard and Hayden 1908; Burrard 1929, 1930). He also presented a proposal to the SoI to make similar changes in their maps. The SoI issued an order to use the term Karakorum instead of the Karakoram adopting the proposal of Burrard. However, based on the recommendation of the permanent committee on the geographical names of RGS, SoI rescinded the order to use the term Karakorum instead of Karakoram (Burrard and Reynolds 1935). This spelling was also used in the context of the long-term Pakistan-German research project Culture Area Karakorum (CAK) during the 1990s (Stellrecht 1998).

The SoI began to survey the Karakoram and Himalaya under the GTS (Godwin-Austen 1864). The nomenclature and alignments on SoI maps during the second half of the nineteenth century were mainly based on the work of Godwin-Austen and his colleagues (Mason 1928). The diagram of the mountain ranges of the Himalayan and Trans-Himalayan system by Burrard and Hayden (1908) is also on similar lines (Anonymous 1929). They presented Kailas Range to the west of the Nubra–Shyok junction that runs parallel to its northern neighbor, the Karakoram Range (Supplementary Fig. 8.4). This regional classification was followed for almost two decades. Availability of limited survey work by the SoI at the high altitude part of the Karakoram Range is attributed to instructions on not to waste time surveying the barren land above an altitude of 15,000 feet (~4572 m) during the second half of the nineteenth century (Mason 1930a).

A long debate started when Mason (1928) proposed fundamental changes in the existing nomenclature of the Karakoram region during the survey of the Shaksgam Valley and Aghil Ranges, which was published in the records of the SoI. Mason (1928) realized that the existing classification of ranges, topography, extent, and nomenclature on early reconnaissance maps of the SoI needs to be updated and required substantial revision using modern methods and equipment, including terrestrial photogrammetry. Mason observed that the name Karakoram was not universally applied to the main Karakoram Range (i.e., Muztagh-Karakoram). Some explorers used Eastern Karakoram, Western Karakoram, Central Karakoram, Karakoram, and even Muztagh for the same region (e.g., Dainelli 1932a, b; Workman and Workman 1917; Trinkler 1930). Neither the explorers nor the surveyors brought out any names for the ranges, but continued to use the term "Karakoram" loosely (Mason 1930b). Therefore, Mason (1930a, b) proposed standardized maps to improve regional classification of ranges, their spatial extent and nomenclature (Supplementary Fig. 8.5). Modern large-scale maps were also required to know the extent of rivers that were flowing in British India (Burrard and Hayden 1908), and to discover ancient routes across the Karakoram for trade purposes (Dainelli 1932a; Younghusband 1930).

Based on Mason's proposal, the Surveyor-General of India, Edward Tandy, mentioned in the preface of the published report in the Records of the Survey of India, dated 17 November 1928,

> Our department should not decide on questions which depend so much on international usage (Mason 1928 pp iii–iv). We can only assist by publishing the suggestions of our best experts, and then hope the RGS, which includes all the principal geographers and explorers interested, will find an early opportunity of discussing the suggestions and arriving at decisions, in which case we shall be happy to accept them and incorporate them on our future maps (Mason 1928 pp iv).

The discussion during the period between 1930 and 1938 on the Karakoram term was mainly presented in the Geographical Journal of the RGS. The editor of the Geographical Journal began the discussion on the term Karakoram and reported that the maps of the mountain systems of the Himalaya and neighboring ranges of India made by the SoI adopted consultation with geologists (Anonymous 1929). The main objective of this paper from 1929 is to advise, "geologists and meteorologists and all who are interested in small-scale maps to consider the questions on the Karakoram term and its extent before they accept the new geographical representations" (Close et al. 1930, pp. 158). The Geographical Journal offered Sidney

Table 8.2 Name of the Karakoram and surrounding ranges proposed by Mason (1928). *Source* Burrard (1929)

	Existing names in the 1920s	Dual name proposed by Mason (1928)
The main range	Karakoram	Muztagh-Karakoram
The mountain region	Karakoram	Karakoram-Himalaya
Outer subsidiary range	Aghil	Aghil-Karakoram
Inner subsidiary range	Kailas	Kailas-Karakoram

Burrard, former Surveyor-General of India, to place his views before the RGS made an authoritative statement against the changes proposed by Mason (Anonymous 1929). Burrard (1929) defended the existing nomenclature of the Karakoram, which was mainly based on his earlier work (i.e., Burrard and Hayden 1908) (Table 8.2; Supplementary Fig. 8.4). He highlighted that the proposed names were too long with the dual name (e.g., Muztagh-Karakoram) and argued about their purpose for science or school maps. Burrard reported that many other high ranges in Asia are also named as Muztagh (e.g., Muztagh Ata in the Pamir) and suggested that the Karakoram term be used for the main range. He also reported that SoI maps of Tibet are purely topographical and no geological survey was used (Burrard 1930). However, some scholars suggested that the nomenclature of the Karakoram must be sought as a compromise between the views of geographers and geologists (Visser 1930). As topographical evidence was incomplete, Mason proposed a detailed geological and geographical investigation for further discussion of nomenclature (Mason 1930a).

During a RGS meeting held under the chairmanship of Charles Close on 12th May 1930, Mason (1930b) presented his paper on *Nomenclature in the Karakoram* and highlighted the contributions of Strachey, Godwin-Austen, Hayward, and Younghusband for using the Muztagh term. He explained its origin as, where Turki is spoken; there are other Muztaghs, just as there are other Karakorams. He also justified the use of the proposed term Muztagh-Karakoram as like the Himalayan Nomenclature (e.g., Nepal Himalaya, Punjab Himalaya) used by Burrard and Hayden (1908). A long discussion on the issue followed this paper. Some scholars (e.g., Younghusband, Tombazi) supported the views of Mason, whereas others strongly contested it (e.g., Burrard). Burrard reported the contributions of Moorcroft, Cunningham, Montgomerie, for using the term Karakoram. Therefore, he postulated for the sake of historical memory that the Karakoram term must be preserved (Bhattacharji 2010; Burrard 1929).

In order to arrive at a consensus on the nomenclature of the Karakoram with the support of Mason's map, a conference was proposed by Charles Close (Mason, 1930b). It was also suggested that RGS should publish a map of the Karakoram to cater to the needs of worldwide interest in the subject. C. G. Lewis, the then Director of the Geodetic Branch of the SoI, was identified to defend the views of the SoI at the conference. After nine informal pre-conference meetings organized by RGS, based on the maps (Fig. 8.4) and blueprints of the surveys a point-by-point draft memorandum was brought out. Burrard, who had taken an interest in the subject, appreciated the committee's effort to address the long (~50 years) outstanding problem of the Karakoram nomenclature (Mason 1938). The official final meeting at a conference held on March 23, 1937 recommended the following extent and boundary of the Karakoram. The Great Karakoram was divided into seven sub-divisions and the lesser Karakoram into five ranges and groups (Mason 1938) (Table 8.3).

> The 1938 Karakoram conference recommended the term "The Karakoram" On the south: by the Shyok river from its bend at about longitude 78° 15' (Survey of India map sheets 52 J, F, B, A, 43 M) to its junction with the Indus, about longitude 75° 55'; then by the Indus to its junction with the Gilgit river about longitude 74° 40' (43 I); and by the Gilgit river (43 1,42 L, H) to the confluence of the Ishkoman river about longitude 73° 45'. On the west: by the Ishkoman and Karumbar rivers (42 H, L) to the Chilinji pass. On the north: from the Chilinji pass, down the Chapursan river, over the Kermin pass to Rich, and down the Kilik river to its junction with the Khun-jerab (42 L); then up the Khunjerab river to the Khunjerab pass, across the head of the Oprang Pamir to the Oprang pass, and down the Oprang river to its junction with the Shaksgam (42 P); then up the Shaksgam river to its source at Wood's Pass "G" (for which we propose the name Shaksgam pass) (42 P, 51 D, 52, A, E); then to the snout of the Rimo-Yarkand river source, and by the left bank of the Rimo glacier to the junction of the Rimo river and the Chip-chap (52 E). On the east: by the upper Shyok from the Rimo-Chip-chap junction to the great bend in the river about longitude 78° 15' (52 E, F, J) (Mason 1938 , 9).

The 1938 Karakoram conference of the RGS excluded both the Aghil Mountains and the Ladakh Range from the Karakoram nomenclature. The conference also recommended the separation of the Karakoram from the Himalayas as two separate mountain systems. A committee appointed by the council of the RGS identified Batchelor, the senior draughtsman to work under the supervision of Mason to produce a map of the Karakoram on the scale of 1:750,000 in 1939 (Anonymous 1940). The names were modified and updated on the map in accordance with the decisions of the 1938 conference and subsequently approved by the Surveyor General of India. However, all seven sub-divisions and the boundary of the Karakoram approved in the conference did not appear in the generalized RGS map. Some portion of the Aghil Mountains and Depsang Plains were also shown as unexplored in this map.

After the partition of British India, the contested boundary claims between the nation-states India, Pakistan, and China complicated accessibility for field research in the Karakoram due to restrictions in parts of Ladakh (controlled by India), Gilgit-Baltistan (controlled by Pakistan), and Shaksgam (controlled by China). While a lot of original maps and data from exploration surveys exist from the colonial period (Hewitt 2008), less material is available for the post-colonial period due to contested territorial claims and political conflicts over the mountain territory. During the 1950s, many large-scale maps of the Karakoram were generated by Italian explorers and mountaineering expeditions (Table 8.1). The first detailed large-scale (1:250,000) orographical sketch map of Karakoram was published by Jerzy Wala in 1976. Since then, he has prepared and updated several large-scale regional maps of the Karakoram Range using satellite data (Table 8.1), including two sheets at a scale of 1:250,000 (Wala 1990). However, overlay of these maps on Landsat orthorectified

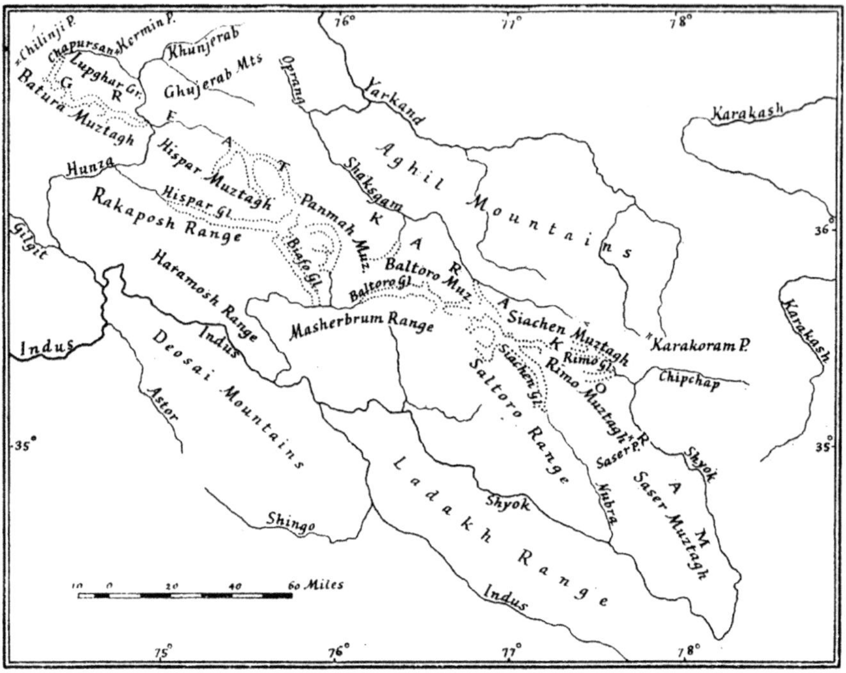

Fig. 8.4 Ranges and extent of the Karakoram according to 1937 conference. *Source* Mason (1938). Map published with the permission of the Royal Geographical Society (with IBG)

Table 8.3 Divisions of Karakoram and its ranges (*Source* Mason 1938)

Primary division	Secondary division	Subgroups
I Great Karakoram	1. Batura Muztagh	
	2. Hispar Muztagh	
	3. Panmah Muztagh	
	4. Baltoro Muztagh	
	5. Siachen Muztagh	
	6. Rimo Muztagh	
	7. Saser Muztagh	
II Lesser Karakoram	1. North of the Great Karakoram	1a. Lupghar Group
		1b. Ghujerab Mountains
	2. Rakaposhi Range	
	3. Haramosh Range	
	4. Masherbrum Range	
	5. Saltoro Range	

satellite images exhibits significant planimetric shift (Fig. 8.1). RGS organized an international conference on the Karakoram to celebrate its 150th anniversary and published two volumes of scientific papers on several research areas such as survey, glaciology, geomorphology, and natural hazards (Miller 1984a, b). In the second volume, Middleton (1984) documented a historical exploration of Karakoram and mentioned the remarkable contribution of many explorers (e.g., Moorcroft, Vigne, Hayward, Younghusband). In the same volume, Tahirkheli and Jan, (1981) reported that the western and eastern boundaries of the Karakoram along the Hindukush and Himalaya, respectively, remain an unsettled issue. They also proposed to shift the western boundary of the Karakoram further west up to the Yasin valley and the eastern boundary up to the Deosai-Ladakh Range. However, this study neither reports the background of the problem nor mentions previous efforts, which are compiled in Table 8.1. Some noteworthy studies (e.g., Zongtai 1988; Guo et al. 2015) also mapped glaciers of the Karakoram during the inventory of the Chinese glaciers.

8.4 Discussion and Conclusions

Explorers from various countries and institutions (e.g., SoI, RGS, Italian Geographical Society, Dutch Geographical Society) have notably contributed to exploration and mapping of the Karakoram Range since the nineteenth century (Table 8.1; Fig. 8.5). For the colonial period, the role of indigenous explorers in field surveys was largely neglected and needs to be studied more thoroughly, as exemplified for the Hunza-Karakoram by Kreutzmann (2020) and for the adjoining Wakhan by Kreutzmann (2017). We observed a gradual change in the spatial extent of the Karakoram during the past three centuries based on the analysis of more than 100 maps of the Karakoram and the surrounding regions. Moorcroft was the first western geographer who introduced the term "Karakoram Mountains" into modern geography during the 1820s. A debate on the Karakoram term began in the 1920s-1930s and consensus was formed by RGS and SoI on its spatial extent during the 1938 Karakoram conference (Mason 1938). Burrard and Mason from the SoI independently made efforts to standardize the regional classification of ranges, spatial extent, and nomenclature of the Karakoram.

Some scholars have delineated outlines on maps based on the description of the 1938 Karakoram conference. Desio (1991) agrees to the southern, northern, and western delineated boundary of the Karakoram, which is mainly bounded by Shyok, Gilgit, Ishkuman, Karumbar, and Shaksgam Rivers. Some uncertainty existed in the NW and NE direction, which is related to passes and short drainage lines. We also found significant subjectivity in boundary mapping while joining the outline of rivers to passes. Desio (1991) suggested that the Karakoram Pass should be included in the extent of the Karakoram Range. Conversely, we delineated the outline of the Karakoram as recommended by the RGS conference on the Karakoram nomenclature (Mason 1938).

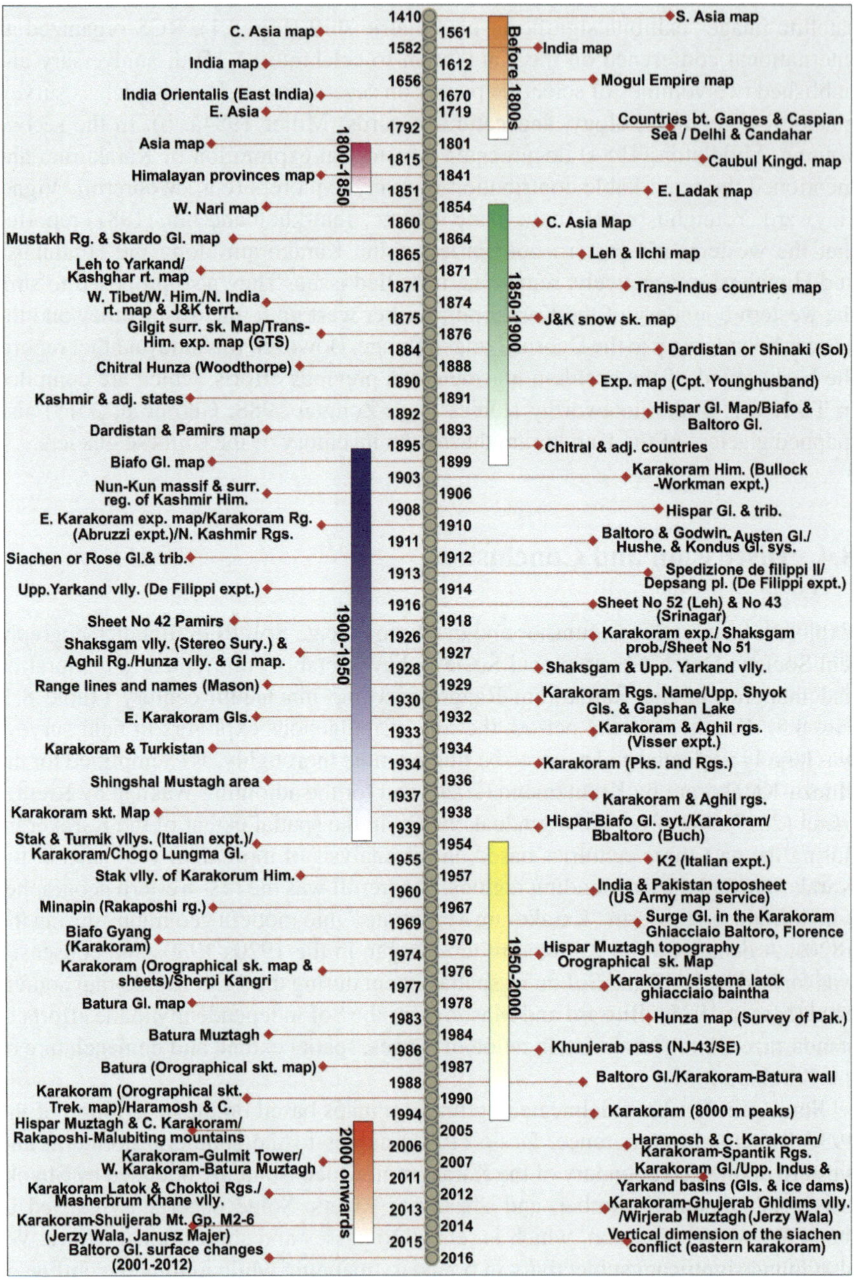

Fig. 8.5 Timeline history of the Karakoram toponymy, its spatial extent and mapping

Table 8.4 Area of the Karakoram boundary in different studies

Studies	Area of Karakoram (km^2)
Bolch et al. (2012)	117,883.3
RGI Consortium (2017)	104,010.4
Rankl et al. (2014)	79,606.7
Wala (1990)	43,045.5
Present study	44,546.0

Literature review suggests that geologists as members of the survey teams were included in most of the Karakoram expeditions during the first half of the twentieth century (e.g., Shipton et al. 1938). The Director of the Geological Survey of India was also consulted for the 1938 nomenclature of the Karakoram conference (Mason 1938). However, Gaetani (1997) reported 'geographical' definition (Mason 1938) hardly coincides with the possible boundary of the 'geological' Karakoram. We also documented several regional geological maps but could not find a complete geological map of the entire Karakoram with its boundary in historical as well as recent literature (Supplementary Table 8.1).

During the current decade (2010s), an inconsistency was observed in the spatial extent of the Karakoram in many recent studies due to the unavailability of a concise digital outline of the mountain range based on the recommendations of the 1938 Karakoram conference. Our estimated area of the Karakoram based on the 1938 Karakoram conference is almost similar (+3%) to the Wala (1990) estimates of the area (Table 8.4). A slight difference (+3%) is possibly due to the extension of the Karakoram to the Khora Bhurt Pass in the northwest instead of Chillinji Pass and further extension of the northeast boundary to the Karakoram pass in the Wala (1990) map (Fig. 8.1). However, the Karakoram extent by Bolch et al. (2012) and RGI Consortium (2017) is ~2.7 and ~2.3 times higher than ours and Wala (1990) estimates, respectively. This large difference in the boundary area estimates can be attributed to the inclusion of surrounding areas like the Hindu Raj, Nanga Parbat, Pamir, Wakhan, Aghil, Chang Chenmo, and Ladakh mountains in the Karakoram region in these studies (Fig. 8.1). Efforts should be made to overcome spatial inconsistencies in future scientific studies and to generate comparable spatial data sets, especially for quantitative analyses. In order to support this goal, we propose the digital outline of the Karakoram extent generated in the present study. The purpose of our study is not to downplay previous efforts rather to generate and provide a widely accepted digital outline of the Karakoram in the public domain based on the 1938 Karakoram conference to enhance clarity and reduce uncertainty in the interest of common understanding. We provide the outline of Karakoram in digital GIS format in a public domain based on the definition given by the RGS and SoI during the 1938 Karakoram conference from the given web link i.e., https://bit.ly/3EGhU67 or directly email to corresponding author.

Acknowledgements The Wadia Institute of Himalayan Geology and Center for Glaciology of the Department of Science and Technology (DST), New Delhi is thankfully acknowledged for financial and infrastructural support. PC is thankful to the Central University of Punjab, Bathinda (India) for providing the research facilities and the RSM grant to support this research. AKG thanks DST, New Delhi for support under J.C. Bose Fellowship (SR-S2/JCB-80/2011) by the Department of Science and Technology, Government of India. We also thank Priyeshu Srivastava for valuable suggestions, which improved the manuscript. RB thanks Tobias Bolch for providing an outline of the Karakoram, Wanqin Guo for discussion on the second Chinese glacier inventory, Dan Jantzen for providing the RGS Karakoram map 1939, and Michael Zemp for discussion on the Randolph Glacier Inventory (and earlier work). We extend our gratitude to Shekhar Pathak and his team for launching and feeding the Pahar website (http://pahar.in/) that provides valuable literature and historical maps on the Karakoram in digital format used in the present study.

References

Anonymous (1929) Mountain names on the Indian border. Geogr J 74(3):274–277. https://doi.org/10.2307/1784366

Anonymous (1940) Note on the construction of the Karakoram map to be published early in Apri9/20/2022l. Geogr J 95(4):299–300. https://doi.org/10.2307/1788466

Archer D (2003) Contrasting hydrological regimes in the upper Indus Basin. J Hydrol 274(1–4):198–210. https://doi.org/10.1016/S0022-1694(02)00414-6

Atkinson TW (1860) Oriental and Western Siberia: a narrative of seven years' explorations and adventures in Siberia, Mongolia, the Kirghis Steppes, Chinese Tartary, and Part of Central Asia. JW Bradley, Philadelphia, USA, p 483

Baghel R, Nüsser M (2015) Securing the heights: the vertical dimension of the Siachen conflict between India and Pakistan in the Eastern Karakoram. Polit Geogr 48:24–36. https://doi.org/10.1016/j.polgeo.2015.05.001

Barrand NE, Murray T (2006) Multivariate controls on the incidence of glacier surging in the Karakoram Himalaya. Arct Antarct Alp Res 38(4):489–498. https://doi.org/10.1657/1523-0430(2006)38[489:MCOTIO]2.0.CO;2

Baume LC (1990) Karakoram maps: trekking routes with short descriptions, peaks, and physical topography. [England]: LeomannMaps

Bhambri R, Bolch T, Kawishwar P, Dobhal DP, Srivastava D, Pratap B (2013) Heterogeneity in glacier response in the upper Shyok valley, northeast Karakoram. Cryosphere 7:1385–1398. https://doi.org/10.5194/tc-7-1385-2013

Bhambri R, Hewitt K, Kawishwar P, Pratap B (2017) Surge-type and surge-modified glaciers in the Karakoram. Sci Rep 7(1):15391. https://doi.org/10.1038/s41598-017-15473-8

Bhattacharji S (2010) Naming the Karakoram: the Debate of the 1930s'. In: Besson F (eds) Mountains figured and disfigured in the english-speaking World. Cambridge Scholars Publishing, Newcastle upon Tyne, UK, pp 531–544

Bolch T, Kulkarni A, Kääb A, Huggel C, Paul F, Cogley JG, Frey H, Kargel JS, Fujita K, Scheel M, Bajracharya S, Stoffel M (2012) The state and fate of Himalayan glaciers. Science 336(6079):310–314. https://doi.org/10.1126/science.1215828

Bolch T, Pieczonka T, Mukherjee K, Shea J (2017) Brief communication: glaciers in the Hunza catchment (Karakoram) have been nearly in balance since the 1970s. Cryosphere 11(1):531–539. https://doi.org/10.5194/tc-11-531-2017

Brescius M, Kaiser F, Kleidt S (eds) (2015) Über den Himalaya. Die Expedition der Brüder Schlagintweit nach Indien und Zentralasien 1854–1858. Köln, Weimar, Wien

Burrard S (1929) The mountains of the Karakoram: a defence of the existing nomenclature. Geogr J 74(3):277–284. https://doi.org/10.2307/1784367

Burrard S (1930) The Himalayan Ranges and Godwin Austen's map. Geogr J 75(1):35–37. https://doi.org/10.2307/1783750

Burrard SG, Reynolds JH (1935) The name Karakoram and reasons for recommending a change in its spelling. Empire Surv Rev 3(16):115–120. https://doi.org/10.1179/sre.1935.3.16.115
Burrard SG, Hayden HH (1908) A sketch of the geography and geology of the Himalaya Mountains and Tibet, vol 1. Superintendent Government Printing, Calcutta, India, p 308
Close C, Burrard S, Younghusband F, Longstaff TG, Ryder CHD, Bruce CG, Neame P, Hinks, Reynolds JH, Tombazi, Mason K (1930) Nomenclature in the Karakoram: discussion. Geogr J 76(2):148–158. https://doi.org/10.2307/1783980
Cogley JG (2011) Present and future states of Himalaya and Karakoram glaciers. Ann Glaciol 52(59):69–73. https://doi.org/10.3189/172756411799096277
Cogley JG (2012) Glaciology: no ice lost in the Karakoram. Nat Geosci 5(5):305–306. https://doi.org/10.1038/ngeo1456
Conway WM (1894) Climbing and exploration in the Karakoram-Himalayas. Nature 50(1287):199–201. https://doi.org/10.1038/050199c0
Dainelli G (1932a) Italia pass in the Eastern Karakoram. Geogr Rev 22(3):392–402. https://doi.org/10.2307/208969
Dainelli G (1932b) A journey to the glaciers of the Eastern Karakoram. Geogr J 74(4):257–274. https://doi.org/10.2307/1784325
De Filippi F (1915) Expedition to the Karakoram and Central Asia, 1913–1914. Geogr J 46(2):85–99. https://doi.org/10.2307/1780168
De Filippi F (1911) The expedition of H. R. H. the Duke of the Abruzzi to the Karakoram Himalayas. Geograph J 37(1):19–25. https://doi.org/10.2307/1777575
De Filippi F (1924) Storia della Spedizione scientifica italiananel Himàlaia, Caracorùm eTurchestàn Cinese (1913–1914), vol 1. N. Zanichelli, Bologna, Italy, p 541
De Filippi F (1932) Italian Expedition to the Himalaya, Karakoram and Eastern Turkestan (1913–1914). Edward Arnold, London
Desio A (1991) Geographical features of the Karakorum. Ismeo, Milano, Italy
Drew F (1875) The Jummoo and Kashmir territories: a geographical account. E. Stanford, London, UK
Dyhrenfurth GO (1955) To the third pole. The history of the High Himalaya. Werner Laurie, London, UK
Elphinstone M (1815) Account of the Kingdom of Caubul, and its dependencies in Persia, Tartary, and India; comprising a view of the Afghaun Nation, and a History of the Dooraunee Monarchy. Cambridge University Press Longman, London UK
Finsterwalder R (1996) Accompanying text for the "Hunza-Karakorum 1: 100000" map. Erdkunde 50(3):169–172. https://doi.org/10.3112/erdkunde.1996.03.01
Forsythe N, Fowler HJ, Li XF, Blenkinsop S, Pritchard D (2017) Karakoram temperature and glacial melt driven by regional atmospheric circulation variability. Nat Clim Change 7(9):664–670
Gaetani M (1997) The Karakorum block in Central Asia, from Ordovician to Cretaceous. Sed Geol 109(3–4):339–359. https://doi.org/10.1016/S0037-0738(96)00068-1
Gardelle J, Berthier E, Arnaud Y (2012) Slight mass gain of Karakoram glaciers in the early twenty-first century. Nat Geosci 5(5):322–325. https://doi.org/10.1038/ngeo1450
Gardelle J, Berthier E, Arnaud Y, Kääb A (2013) Region-wide glacier mass balances over the Pamir-Karakoram-Himalaya during 1999–2011. Cryosphere 7(4):1263–1286. https://doi.org/10.5194/tc-7-1263-2013
Gibson MJ, Glasser NF, Quincey DJ, Rowan AV, Irvine-Fynn TD (2017) Changes in glacier surface cover on Baltoro glacier, Karakoram, north Pakistan, 2001–2012. J Maps 13(2):100–108. https://doi.org/10.1080/17445647.2016.1264319
Godwin-Austen HH (1864) On the glaciers of the Mustakh range. J R Geogr Soc Lond 34:19–56. https://doi.org/10.2307/1798464
Godwin-Austen HH, Conway WM (1893a) The Nomenclature of the Karakoram Peaks. Geogr J 1(4):368–371. https://doi.org/10.2307/1773906
Godwin-Austen HH, Conway WM (1893b) The nomenclature of the Karakoram Peaks. Geogr J 1(2):177–178. https://doi.org/10.2307/1773762

Groos AR, Mayer C, Smiraglia C, Diolaiuti G, Lambrecht A (2017) A first attempt to model region-wide glacier surface mass balances in the Karakoram: findings and future challenges. Geogr Fis Din Quat 40(2):137–159. https://doi.org/10.4461/GFDQ.2017.40.10

Gunn JP (1930) Report on the Chong Kumdan dam and Shyok flood of 1929. Public Work Department, Punjab Government, p 29

Guo W, Liu S, Xu J, Donghui S, Xiaojun Y, Junfeng W, Weijia B, Pengchun Y, Qiao L, Zongli J (2015) The second Chinese glacier inventory: data, methods and results. J Glaciol 61(226):357–372. https://doi.org/10.3189/2015JoG14J209

Hayward GW (1870) Journey from Leh to Yarkand and Kashgar, and exploration of the sources of the Yarkand River. J R Geogr Soc Lond 40(1):41–74. https://doi.org/10.2307/1799609

Hedin SA (1922) Southern Tibet: discoveries in former times compared with my own researches in 1906–1908, vol 7. Lithographic institute of the General staff of the Swedish army, Stockholm, Sweden

Heichel W (2010) Chronik der Erschließung des Karakorum, Teil II - Central Karakorum I Neubiberg. Universität der Bundeswehr München, Germany, p 460

Hewitt K (1969) Glacier surges in the Karakoram Himalaya (Central Asia). Can J Earth Sci 6(4):1009–1018. https://doi.org/10.1139/e69-106

Hewitt K (2005) The Karakoram anomaly? Glacier expansion and the 'elevation effect,' Karakoram Himalaya. Mount Res Dev 25(4):332–340. https://doi.org/10.1659/0276-4741(2005)025[0332:TKAGEA]2.0.CO;2

Hewitt K, Liu J (2010) Ice-dammed lakes and outburst floods, Karakoram Himalaya: historical perspectives on emerging threats. Phys Geogr 31(6):528–551. https://doi.org/10.2747/0272-3646.31.6.528

Hewitt K (2008) Rediscovering colonised landscapes: the first Europeans at the Mustagh Pass, Karakoram Himalaya, Inner Asia. In: Gervers M, Bulag U, Long G (eds) The exploitation of the landscape of central and inner Asia (Toronto Studies in Central and Inner Asia, 9). Toronto

Institute of Glaciology (1978) The map of Bhatura Glacier. Cryopedology and Desert Research Academia Sinica Lanchow, China

Iturrizaga L (2011) Trends in 20th century and recent glacier fluctuations in the Karakoram Mountains. Zeitschrift Für Geomorphol Suppl Issue 55(3):205–231. https://doi.org/10.1127/0372-8854/2011/0055S3-0059

Johnson WH (1867) Report on his journey to Ilchi, the Capital of Khotan, in Chinese Tartary. J R Geogr Soc Lond 37:1–47. https://doi.org/10.2307/1798517

Kapnick SB, Delworth TL, Ashfaq M, Malyshev S, Milly PC (2014) Snowfall less sensitive to warming in Karakoram than in Himalayas due to a unique seasonal cycle. Nat Geosci 7(11):834–840

Kääb A, Berthier E, Nuth C, Gardelle J, Arnaud Y (2012) Contrasting patterns of early twenty-first-century glacier mass change in the Himalayas. Nature 488(7412):495–498. https://doi.org/10.1038/nature11324

Kääb A, Treichler D, Nuth C, Berthier E (2015) Brief communication: contending estimates of 2003–2008 glacier mass balance over the Pamir-Karakoram-Himalaya. Cryosphere 9(2):557–564. https://doi.org/10.5194/tc-9-557-2015

Khan A, Richards KS, Parker GT, Mcrobie A, Mukhopadhyay B (2014) How large is the Upper Indus Basin? The pitfalls of auto-delineation using DEMs. J Hydrol 509:442–453. https://doi.org/10.1016/J.JHYDROL.2013.11.028

Kick W (1956) Chogo Lungma Glacier, Karakoram. Geogr J 122(1):93–96. https://doi.org/10.2307/1791480

Kick W (1960) The first glaciologists in Central Asia. J Glaciol 3(28):686–692. https://doi.org/10.3189/S0022143000017974

Knight EF (1893) Where three empires meet: a narrative of recent travel in Kashmir, Western Tibet, Gilgit and the Adjoining Countries. Longman Green, London, UK

Kreutzmann H (2007) Geographical Research in Chinese Central Asia: aims and ambitions of international explorers in the 19th and 20th centuries. Die Erde 138(4):369–384

Kreutzmann H (2017) Wakhan Quadrangle: exploration and espionage during and after the great game. Harrasowitz, Wiesbaden, Germany
Kreutzmann H (2020) Hunza matters: ordering and bordering between ancient and new Silk Roads. Harrasowitz, Wiesbaden, Germany
Li XF, Fowler HJ, Forsythe N, Blenkinsop S, Pritchard D (2018) The Karakoram/Western Tibetan vortex: seasonal and year-to-year variability. Clim Dyn 51(9):3883–3906
Longstaff TG (1910) Glacier exploration in the Eastern Karakoram. Geogr J 35(6):622–653. https://doi.org/10.2307/1777235
Marsh C (1876) Description of a trip to the gilgit valley, dependency of the Maharaja of Kashmir. J Asiatic Soc Bengal 45(1):119–130
Mason K (1927a) The Shaksgam valley and Aghil range. Geogr J 69(4):289–323. https://doi.org/10.2307/1782759
Mason K (1927b) The stereographic survey of the Shaksgam. Geogr J 70(4):342–352. https://doi.org/10.2307/1782060
Mason K (1930a) The proposed nomenclature of the Karakoram-Himalaya. Geogr J 75(1):38–44. https://doi.org/10.2307/1783751
Mason K (1930b) Nomenclature in the Karakoram. Geogr J 76(2):143–148. https://doi.org/10.2307/1783979
Mason K (1938) Karakoram nomenclature. Geogr J 91(2):123–152. https://doi.org/10.2307/1788003
Mason K (1928) Exploration of the Shaksgam Valley and Aghil Ranges, 1926. The Geodetic Branch Office, Survey of India, Dehradun, India, p 182
Mason K (1955) Abode of snow: a history of Himalayan exploration and mountaineering. Rupert Hart-Davis, London, UK
Middleton D (1984) Karakoram history: early exploration. In: Miller K (ed) The International Karakoram Project, vol 1. UK J University Press, Cambridge, pp 17–31
Miller KJ (1984a) The international Karakoram project, vol 1. Cambridge University Press, Cambridge, UK
Miller KJ (1984b) The international Karakoram PROJECT, vol 2. Cambridge University Press, Cambridge, UK
Moorcroft W, Trebeck G (1841) Travels in the Himalayan provinces of Hindustan and the Panjab. In: Ladakh and Kashmir; in Peshawar, Kabul, Kunduz, and Bokhara, vol 1. London, UK
Morris CJ (1928) Some valleys and glaciers in Hunza. Geogr J 71(6):513–531. https://doi.org/10.2307/1783172
Mott PG (1950) Karakoram survey, 1939: a new map. Geogr J 116(1/3):89–95. https://doi.org/10.2307/1789515
Neve A (1910) The ranges of the Karakoram. Geogr J 36(5):571–577. https://doi.org/10.2307/1777345
Paul F (2015) Revealing glacier flow and surge dynamics from animated satellite image sequences: examples from the Karakoram. Cryosphere 9(6):2201–2214. https://doi.org/10.5194/tc-9-2201-2015
Phillimore RH (1945) Historical records of the survey of India, vol 1. In: 18th Century. The Geodetic Branch Office, Survey of India, Dehradun, India, p 400
Phillimore RH (1950) Historical records of the survey of India, vol 2. The Geodetic Branch Office, Survey of India, Dehradun, India
Pratap B, Dobhal DP, Bhambri R, Mehta M, Tewari VC (2016) Four decades of glacier mass balance observations in the Indian Himalaya. Reg Environ Change 16(3):643–658. https://doi.org/10.1007/s10113-015-0791-4
Quincey DJ, Glasser NF, Cook SJ, Luckman A (2015) Heterogeneity in Karakoram glacier surges. J Geophys Res Earth Surf 120(7):1288–1300. https://doi.org/10.1002/2015JF003515
Rankl M, Kienholz C, Braun M (2014) Glacier changes in the Karakoram region mapped by multimission satellite imagery. Cryosphere 8(3):977–989. https://doi.org/10.5194/tc-8-977-2014

Reggiani P, Mukhopadhyay B, Rientjes THM, Khan A (2017) A joint analysis of river runoff and meteorological forcing in the Karakoram, upper Indus Basin. Hydrol Process 31(2):409–430. https://doi.org/10.1002/hyp.11038

Rennell J (1793) Memoir of a map of Hindoostan, or the Mogul empire. Brown M, London, UK, p 295

RGI Consortium (2017) Randolph glacier inventory—a dataset of global glacier outlines: version 6.0: technical report, global land ice measurements from space. Colorado, USA. Digital Media. https://doi.org/10.7265/N5-RGI-60

Schomberg RCF (1936) Unknown Karakoram. M. Hopkinson, Ltd., London, UK

Shaw R (1871) Visits to high tartary, Yârkand, and Kâshghar: formerly Chinese tartary. John Murray, London, UK

Shipton E, Spender M, Auden JB (1938) The Shaksgam expedition, 1937. Geogr J 91(4):313–336. https://doi.org/10.2307/1788187

Shipton E (1938) Blank of the map. Hodder and Stoughton, London, UK, p 299

Stellrecht I (1998) Karakorum-Hindukusch-Himalaya: dynamics of change. In: Culture area Karakorum scientific studies 4/1 and 4/2. Köln

Strachey H (1853) Physical geography of western Tibet. J R Geogr Soc Lond 23(1853):1–69. https://doi.org/10.2307/1797948

Tahirkheli RAK, Jan QM (1981) The geographical and geological domains of the Karakoram. In: Miller KJ (ed) The International Karakoram Project, vol 1. Cambridge University Press, Cambridge, UK, pp 57–70

Trinkler E (1930) The ice-age on the Tibetan Plateau and in the adjacent regions. Geogr J 75(3):225–232. https://doi.org/10.2307/1784006

Ullah M (1843) Travels beyond the Himalaya by Mir Izzet Ullah Republished from the Calcutta Oriental Quarterly Magazine, 1825. J R Asiat Soc Great Britain Ireland 7(2):283–342

Visser PC (1926) Explorations in the Karakoram. Geogr J 68(6):457–468. https://doi.org/10.2307/1782001

Visser PC (1930) The mountains of Central Asia and their nomenclature. Geogr J 76(2):138–142. https://doi.org/10.2307/1783978

Visser PC (1934) The Karakoram and Turkestan expedition of 1929–1930. Geogr J 84(4):281–295. https://doi.org/10.2307/1786691

Visser PC, Visser-Hooft J (1938) Wissenschaftliche Ergebnisse der niederländischen Expeditionen in den Karakorum und die angrenzenden Gebiete in den Jahren 1922, 1925 und 1929–30. E.J. Brill, Leiden, Netherlands, p 212

Wala J (1990) Karakoram orographical sketch map, Scale 1:250 000, Sheet 1, Sheet 2, Zurich

Workman FB (1910) Its tributaries and mountains. Geogr J 35(2):105–115. https://doi.org/10.2307/1776943

Workman WH (1912) Latest Himalayan explorations of the workmans. Bull Am Geogr Soc 44(1):17–19. https://doi.org/10.2307/199738

Workman FB (1914) The exploration of the siachen or rose glacier Eastern Karakoram. Geograph J 43(2):117–141. https://doi.org/10.2307/1778919

Workman FB, Workman WH (1900) In the ice world of Himálaya: among the peaks and passes of Ladakh, Nubra, Suru, and Baltistan. T. Fisher Unwin, London, UK, p 204

Workman FB, Workman WH (1908) Ice-bound heights of the Mustagh: an account of two seasons of pioneer exploration in the Baltistan Himálaya. Charles Scribner's Sons, New York, USA, p 444

Workman FB, Workman WH (1917) Two summers in the ice-wilds of Eastern Karakoram: the exploration of nineteen hundred square miles of mountain and glacier. T. Fisher Unwin, London, UK, p 296

Workman FB (1906) First exploration of the Hoh Lumba and Sosbon Glaciers: two pioneer ascents in the Himalayas. Geograph J 27(2):129–141. https://doi.org/10.2307/1776665

Younghusband F (1895) Chitral, Hunza, and the Hindu Kush. Geogr J 5(5):409–422. https://doi.org/10.2307/1773852

Younghusband F (1926) The problem of the Shaksgam valley. Geogr J 68(3):225–230. https://doi.org/10.2307/1782580

Younghusband F (1930) The Muztagh pass in 1887. Geogr J 76(6):522–524. https://doi.org/10.2307/1783698

Zhou Y, Li Z, Li J (2017) Slight glacier mass loss in the Karakoram region during the 1970s to 2000 revealed by KH-9 images and SRTM DEM. J Glaciol 63(238):331–342. https://doi.org/10.1017/jog.2016.142

Zongtai W (1988) New statistical figures and distribution feature of glaciers on the various mountains in China. Arid Land Geography 11(3):8–14

Chapter 9
Spatial Distribution and Trend Estimation of Tropospheric Formaldehyde: A Space-Borne Observation Over South Asia

Ujjal Deka Baruah and Nitashree Mili

Abstract This chapter deals with the distribution and recent changes of tropospheric formaldehyde as measured from space borne sensors aboard Aura/OMI satellite of 0.1° × 0.1° spatial resolution during 2008–2018 over South Asia. The concentration of formaldehyde is mainly concentrated in the Indo-Gangetic Plains of South Asia spanning Afghanistan, Bangladesh, Bhutan, India, Nepal, Pakistan and Sri Lanka. Thiel-Sen's analysis indicates a decreasing trend (−0.004 $DUyr^{-1}$) during the study period.

Keywords HCHO · Formaldehyde · South Asia · Satellite observation · Trend · Spatial distribution

9.1 Introduction

Formaldehyde (HCHO) is one of the precursors of tropospheric ozone (O_3) formed due to oxidation of hydrocarbons. The major source of HCHO is methane (CH_4), however, non-methane hydrocarbons also act as a principal source over continental atmospheres. Most of the HCHO sinks either due to photolysis or after reacting with the hydroxyl radical (OH). The presence of HOx radicals in the troposphere are formed due to photolysis of HCHO. Formaldehyde is a short-lived tropospheric molecule (~1–5 h) as it photolyses very quickly and reacts readily with OH (Arlander et al. 1995; De Smedt et al. 2008). After photodecomposition of HCHO to HCO, it reacts with oxygen (O_2) to form carbon monoxide (CO). However, photolysis also produces hydroperoxyl radicals (HO_2) that react with nitric oxide (NO) to form NO_2. These atmospheric molecules acts as precursors of ozone (O_3) and nitrogen dioxide (NO_2), which are harmful gases adversely affecting all life forms (David and Nair 2013; Krotkov et al. 2017; Martin et al. 2004; Sinha et al. 2015). Emission of these trace gases affects atmospheric chemistry at global levels (Crutzen and Andreae 1990; Crutzen et al. 1979). Biomass burning, biogenic emissions, insolation, high

U. D. Baruah (✉) · N. Mili
Geography Department, Cotton University, Guwahati, India
e-mail: udbmail@gauhati.ac.in

A. Saikia and P. Thapa (eds.), *Environmental Change in South Asia*,
https://doi.org/10.1007/978-3-030-47660-1_9

temperatures, and anthropogenic emissions are the major sources of O_3, NO_x, and HCHO (Fu et al. 2007; Meyer-Arnek et al. 2005; Yang Wang et al. 2019). While ground-based studies are carried out with higher accuracy, such studies are very limited in nature (Chan et al. 1998; Fishman et al. 2003; Kanaya et al. 2009; Mckendry et al. 1997; Zhang et al. 2008). However, satellite based observations enable us to overcome such limitations (Abbot et al. 2003; De Smedt et al. 2008; De Smedt et al. 2010; Meyer-Arnek et al. 2005).

There are numerous studies based on satellite observations (Abbot et al. 2003; De Smedt et al. 2010; Lieschke et al. 2019; Millet et al. 2008; Spinei et al. 2018). These studies are mostly based on GOME and SCIAMACHY satellites or atmospheric chemistry models (GEOS-CHEM, MOZART-4) (De Smedt et al. 2010; Palmer 2003; Yarragunta et al. 2019). However, we used Aura/OMI satellite derived daily L3 weighted mean tropospheric formaldehyde at $0.1° \times 0.1°$ resolution during 2008–2018 (OMHCHOd) (Chance 2019; David and Nair 2013). We primarily focused our study on spatio–temporal variation of HCHO over South Asia due to its deteriorating air quality caused by increasing population, agriculture expansion, higher energy consumption, urbanization, and industrialization as well as the availability of limited ground measurements (Azad and Kitada 1998; Badarinath et al. 2006; Baek et al. 2014; Gurjar et al. 2008).

9.2 Data and Methods

9.2.1 Satellite Observation

We use the Ozone Monitoring Instrument (OMI) 2008–2018 data recorded by the nadir-looking near-UV/VISCharge Coupled Device (CCD) spectrometer onboard the Aura satellite launched on July 2004 under the Earth Observing System (EOS) program by NASA (Bruni Zani et al. 2020; Chance 2019; Millet et al. 2008; ul-Haq et al. 2015). The product used for the study (OMHCHOd) is a total column amount cloud-screened daily L3 weighted mean tropospheric formaldehyde at $0.1° \times 0.1°$ resolution (Bruni Zani et al. 2020; Sullivan et al. 2016). OMI delivers a wide range of products like O_3 column, SO_2, NO_2, HCHO, BrO, OClO, and aerosols used for air quality measurements. The spectral range covers 264–504 nm with spectral resolution ranging between 0.42 and 0.63 nm. However, OMHCHOd data measurements lie within the spectral range of 327.5–356.5 nm with an uncertainty of 35% (Sullivan et al. 2016). It covers a ground footprint of 13×24 km^2. The higher spectral resolution and increased number of observed wavelengths in OMI is better than the existing GOME, TOMS, and SCIAMACHY satellite observations for air quality and trace gas monitoring and measurements (Chance 2019). Infact, there is also a high correlation between HCHO of OMI and the GEOS-Chem 3D atmospheric model (Baek et al. 2014), whereas correlation is lower between GOME HCHO and GEOS-Chem HCHO (Palmer et al. 2001).

9.2.2 Trend Analysis: Theil Sen' Slope Estimation Method

We used Theil-Sen (TS) slope estimation method to detect the trend (Gilbert 1987; Sen 1968; Theil 1950). This applied method is generally used (Kahya and Kalaycı 2004) to detect linear trends that are not as sensitive to outliers (Van Belle et al. 1984). This method can usually estimate slope with missing data or even with less than 20% measurements records (Helsel and Hirsch 1995) and is widely used to detect the trend of trace gases in atmospheric chemistry (Chen et al. 2020; Toro et al. 2017).

9.2.3 Change Point Detection: Pettitt's Test

Petttitt's test (Pettitt 1979) was applied to detect whether a significant change point occurred in the time series (Jaiswal et al. 2015; Pettitt 1980; Salarijazi 2012). This test is widely used considering its sensitivity to breaks in the middle of a time series (Wijngaard et al. 2003). This change-point analysis also was used to detect any breaks in the time series and to better help interpret the linear trend estimates. If a significant change point exists, the series will be segmented at the change point location into two sub-series.

9.3 Results and Discussion

9.3.1 Spatial Pattern

Figure 9.1 indicates the spatial pattern of averaged annual tropospheric HCHO column over South Asia during 2008–2018. The average annual HCHO column levels ranged between near zero and a maximum of 0.31 DU during 2008–18, higher than during 2005–16(-0.23 DU or 06.22 ± 0.14 molecule cm^{-2}) (Rana et al. 2019). The concentration of HCHO is high along the Indo-Gangetic plains (IGP) in Bangladesh, India, and Pakistan similar to the previous studies (Rana et al. 2019; Yarragunta et al. 2019). HCHO also dominates in the Western Ghats and part of Andhra Pradesh and Telengana, Kerala, and its neighbors due to forest cover releasing higher isoprene emissions (Rana et al. 2019). The major contributors to HCHO sources that lead to higher concentration in these regions are biogenic, thermal power plants, industries, pyrogenic (i.e. crop residue and biomass burnings), and fossil fuel combustions (Baek et al. 2014; Biswas et al. 2019; Mahajan et al. 2015; Rana et al. 2019).

Country-level spatial averages (Fig. 9.2) show that India has the highest concentration annual averaged HCHO levels, while Nepal, Afghanistan, and Bhutan have low levels of HCHO over South Asia. It would appear that variations of HCHO are primarily elevation dependent (Figs. 9.1 and 9.2). Areas with higher elevation viz.

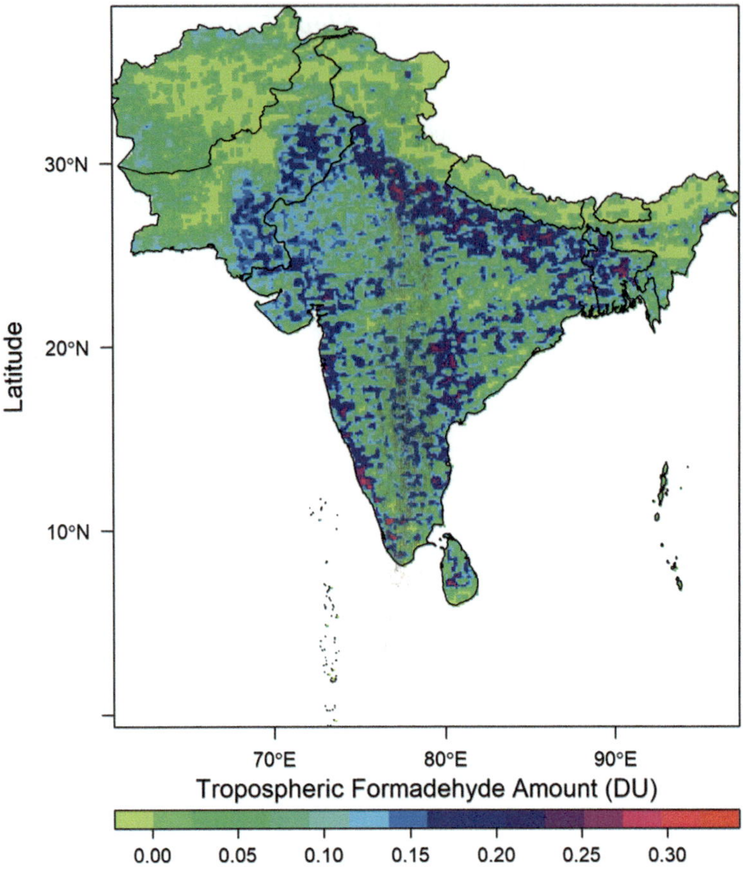

Fig. 9.1 Spatial variations of averaged annual near-surface tropospheric column HCHO, over South Asia. HCHO values were capped between 0 and 2.6×10^{16} molecules/cm^{2}or 1 DU to emphasize feature

Nepal, Bhutan, and Afghanistan have low concentration of HCHO, while lowlands of the Indo-Gangetic Plains (IGP) encompassing Bangladesh, India, and Pakistan possessed relatively high levels of HCHO. The reason behind such increased levels of HCHO in the plains may be the high population density (Boeke et al. 2011). To meet the increasing demands for food, transport, industries of ever-increasing population in the plains, there exists intense use of automobiles, establishment of industries, crop residue and biomass burning. Such activities lead to additional emissions of HCHO apart from natural emissions. Thus, the plains tend to have higher levels of HCHO in comparison to mountainous regions (Biswas et al. 2019; Mahajan et al. 2015).

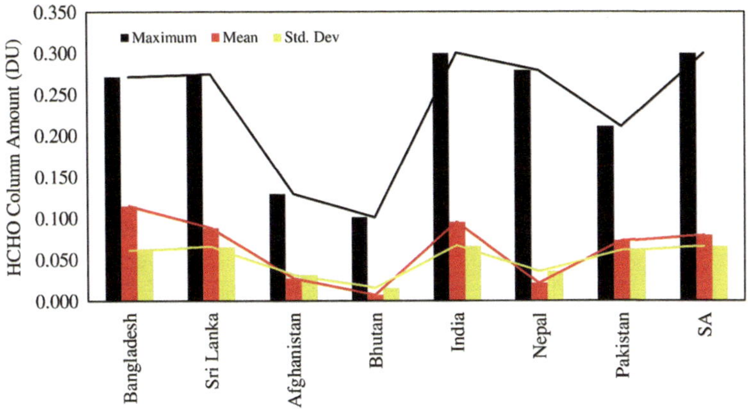

Fig. 9.2 Country-wise levels (mean, maximum, and standard deviation) of HCHO during 2008–18

9.3.2 Seasonal Pattern

Figure 9.3 shows the variation monthly averaged HCHO over South Asia during 2008–2018. Massive emission of HCHO occurs during May–June and October–November in North West India and East Pakistan (Indo-Gangetic Plains) due to the burning of crop residue in Punjab, Uttar Pradesh, and Haryana (Singh and Kaskaoutis 2014; Vadrevu et al. 2011). Usually, farmers prepare their agricultural field for seasonal planting, during the pre-monsoon and post-monsoon season. The months of May–June and October–November are two major agricultural field preparation seasons (harvesting season) in NW India and East Pakistan (Bhuvaneshwari et al. 2019; Singh and Kaskaoutis 2014; Vadrevu et al. 2011). The smoke emitted from burning of crop residues releases organic particulate matter and black carbon (Cusworth et al. 2018), while the harvest season is characterized by post-monsoon season weather conditions. The stagnant northwesterly surface winds favor the travel of smoke several hundred kilometers from its emission source in the IGP regions (Singh and Kaskaoutis 2014).

This crop residue burning is a major constituent of air pollution which adversely affects NW India and East Pakistan (Cusworth et al. 2018; Singh and Kaskaoutis 2014). New Delhi has been a center of discussion for being one of the major air polluted cities along with Ghaziabad, India, and Gurjanwal and Faisalabad in Pakistan of the IGP region (Chhabra et al. 2001; Cusworth et al. 2018). India generates 500 Mt of crop residue every year. Most of the crop residue is used as fodder, energy production, and for industrial purposes. Some 96 Mt is burned out of 140 Mt surplus crop residue (Bhuvaneshwari et al. 2019) with most residue coming from sugarcane, rice, and wheat (Bhuvaneshwari et al. 2019). In addition to the crop residue burning, there is an increase in non-methane volatile organic compounds (NMVOCs) emissions during May–June. NMVOCs like isoprene and monoterpenes are released due to longer sunlight and higher temperature leading to their increased oxidation (MacDonald et al. 2012; Mahajan et al. 2015).

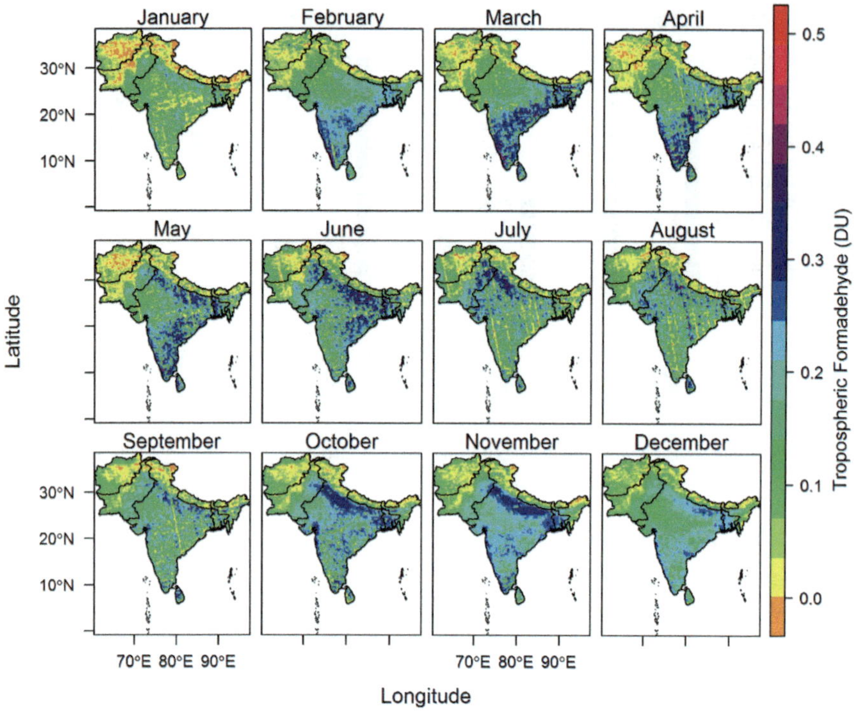

Fig. 9.3 The monthly averaged spatio-temporal variation of tropospheric HCHO over South Asia, 2008–2018

Occasionally, widespread wildfires may occur during drought-like situation years due to dry weather conditions caused by ENSO during pre-monsoon and post-monsoon (Field et al. 2016; Marlier et al. 2013) seasons. Such wildfires enhance the HCHO level in addition to HCHO released due to crop residue burning and biogenic emissions. As far as South East India is concerned, high levels of HCHO were observed during March-June due to biogenic volatile organic compound (VOC) (isoprene) emissions and biomass burning along with burning of the crop residue or cultivation of sugarcane (Fu et al. 2007; Surl et al. 2018). However, a significant peak of HCHO during February–March in North East India and Bangladesh is due to biomass burning (Surl et al. 2018). The lowest level of HCHO concentration was observed during the South West Monsoon period (July–September). The low values of HCHO during summer/monsoon season (July–September) can possibly be attributed to low speed winds, high humidity, and rainfall driven washout out of the harmful trace gases (David and Nair 2013; Lelieveld et al. 2018; Wang et al. 2018). However, the peaks of HCHO column during May–June (Figs. 9.3 and 9.4) cannot be attributed to seasonal variation of rainfall and temperature alone (David and Nair 2013; Wang et al. 2018).

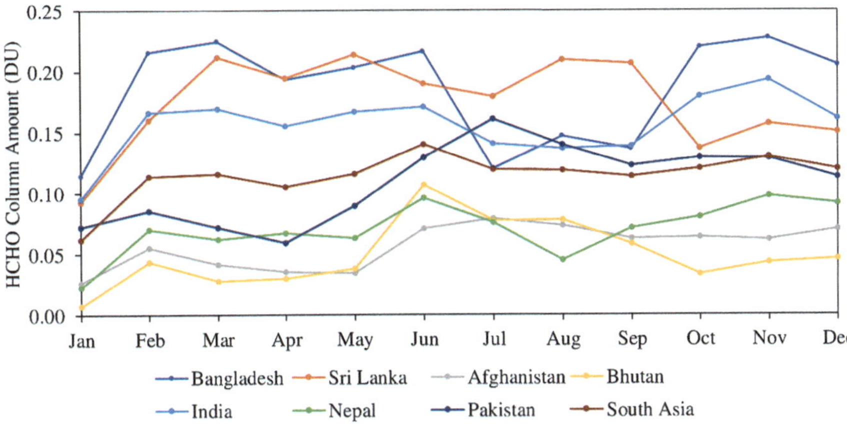

Fig. 9.4 Country-wise variation of average monthly HCHO column over South Asia, 2008–18

All countries follow almost similar seasonal pattern with maximum during May–June and minimum during July–September (Fig. 9.4). Countrywise HCHO maximums were observed in the month of November over Bangladesh (0.23 DU) and minimum over Bhutan (0.01 DU) in January. Sri Lanka surpassed India in terms of the HCHO column, probably due its smaller and more compact geographical area and less topographical and climatic variability. The high temperature and sunlight helps in photochemistry and release of HCHO molecules. Thereafter, India and Pakistan follow as countries of high HCHO column with maximum (0.19 DU) in November and (0.16 DU) in July, respectively. Pakistan and Afghanistan peaked in July instead of June (Fig. 9.4), due to a drier climate than that of India (Chaudhary 1994; Kumar et al. 2010; Salma et al. 2012) following late arrival of SW monsoon (1st July) coupled with anthropogenic emissions (Latif et al. 2016; Lu et al. 2018).

In addition, the biogenic VOC emissions are less than anthropogenic emissions in all mega cities and do not vary much as industrial air pollutants and fuel combustion are the primary source of HCHO emission (Baek et al. 2014).

Figure 9.5 shows the variation of averaged seasonal HCHO over South Asia during 2008–2018. All South Asian countries are dominated by South West Monsoon (Schott and McCreary 2001) and have distinct seasonal cycles i.e. DJF (Winter), MAM (Spring), JJA (Summer), and SON (Autumn) based on astronomical definition of the season (Pielke et al. 1987). The HCHO levels are expected to follow the meteorological cycle based on conventional definition of the season (Pielke et al. 1987). It was evident from the monthly variation of HCHO (Fig. 9.4) that its peaks did not typically follow the meteorological seasonal cycle (Fig. 9.5). It is observed that the HCHO corresponded with the new definition of season, which is based on the dominant synoptic system over the region (Alpert et al. 2004). Over South Asia, the HCHO level peaks in the month of June but subsides during July–August in the same season (summer). Similarly, HCHO level was observed to be low in September but peaked in November during autumn. Therefore, the HCHO classification on the basis of conventional definitions (Fig. 9.5) is not recommendable for seasonal observation of tropospheric HCHO. Here, the use of new definition of seasons may be used for interpretations of cyclic patterns of HCHO. However, the peaks in June and November clearly dominate the summer and autumn seasons (Figs. 9.5 and 9.6) if we go by the astronomical definition of seasons. The seasonal pattern also indicates the spatial distribution pattern of HCHO over South Asia. HCHO dominates South India during summer season and North India specially the Indo-Gangetic Plains (IGP) during autumn (Fig. 9.5).

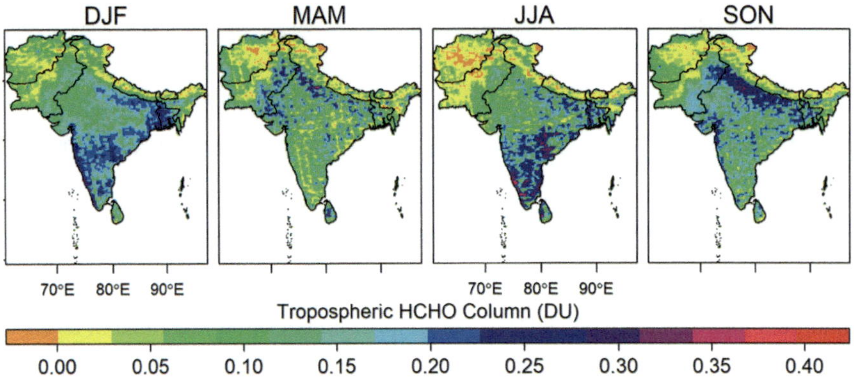

Fig. 9.5 The spatial variation of mean seasonal HCHO over South Asia during 2008–2018

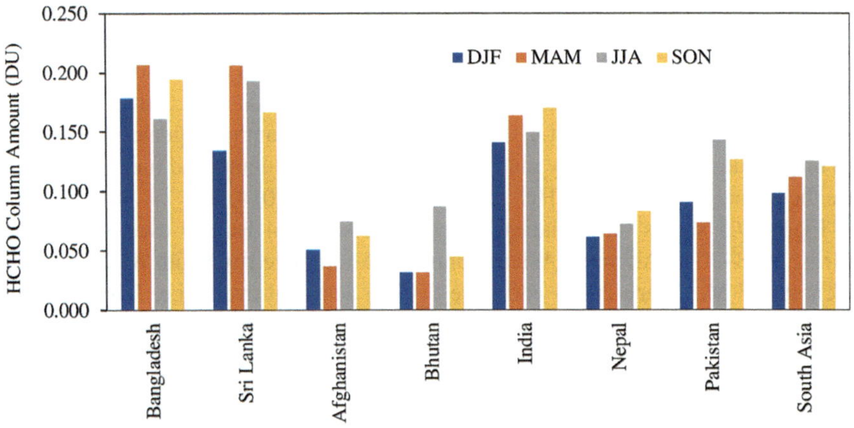

Fig. 9.6 Country-wise seasonal variation of HCHO column over South Asia

Table 9.1 Trend of HCHO over South Asia, 2008–2018

Trace gas	Country	z statistic	Sen's slope (DU Yr^{-1})	95% confidence interval (DU Yr^{-1})		Sen's p-value	Pettitt's	
							Change point (Year)	p-value
HCHO	Bangladesh	−2.024	**−0.0024**	−0.007	−0.0001	0.043	**2011**	0.078
	Sri Lanka	−2.491	**−0.0109**	−0.0151	−0.0022	0.013	2011	0.122
	Afghanistan	−2.803	**−0.0022**	−0.0028	−0.0007	0.005	**2011**	0.078
	Bhutan	−1.557	−0.1195	−0.0051	−0.001	0.120	2012	0.185
	India	−2.958	**−0.0047**	−0.006	−0.0023	0.003	2011	0.185
	Nepal	−2.803	**−0.0051**	−0.0067	−0.002	0.005	2012	0.185
	Pakistan	−1.713	**−0.0016**	−0.004	−0.0003	0.087	2011	0.185
	South Asia	−3.114	**−0.0641**	−0.0048	−0.0022	0.002	2011	0.122

Significant at 0.1 level of significance

9.3.3 HCHO Trend

Over South Asia, we found a decreasing HCHO trend of −0.0641 DU yr^{-1} during 2008–2018 (Table 9.1 and Fig. 9.7). Similar decreasing trends were observed, for all the countries within South Asia, in contrast to increasing trends observed in earlier studies during 1997–2009 (De Smedt et al. 2010). Except Bhutan, all other countries of South Asia show a significant decreasing trend (significance level at 0.1). The highest decreasing trend value of HCHO was found to be −0.01 DU Yr^{-1}over Sri Lanka followed by Nepal (−0.0051 DU Yr^{-1}). Additionally, a significant change point was detected over Afghanistan and Bangladesh in the year 2011.

Figure 9.9 (right) shows the country-wise location of highest increasing and decreasing trends of HCHO over South Asia. The high altitude locations in India and Afghanistan show higher increasing trend even greater than the biomass burning and anthropogenic emissions areas. The only explanation that may be attributed in this context is that of the south facing slopes of the high altitude locations. These tend to receive sunlight for a longer span, and this enhances the photolysis process in releasing HCHO. Additionally, non-dominance of SW monsoon ensures no or less rain for a washout of HCHO. The highest increasing trend value was found to be 0.024 DU Yr^{-1}, located in India followed by Pakistan (0.022DU Yr^{-1}) (Figs. 9.8 and 9.9). The highest decreasing trend value was observed in India (−0.035 DU Yr^{-1}), followed by Sri Lanka (−0.027 DU Yr^{-1}) (Figs. 9.8 and 9.9).

9.4 Conclusion

We studied the spatial distribution pattern and estimated trend of HCHO over South Asia during 2008–2018. For the purpose, we used $0.1° \times 0.1°$ resolution OMI gridded

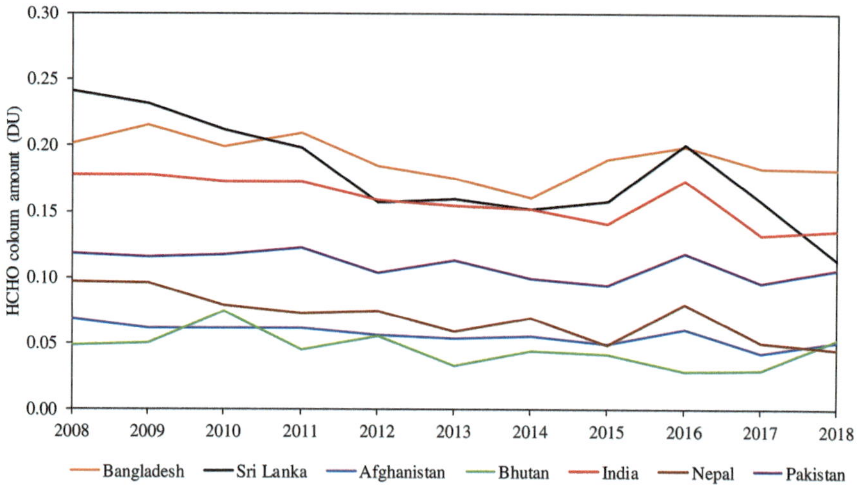

Fig. 9.7 Country wise trend line of HCHO over South Asia, 2008–2020

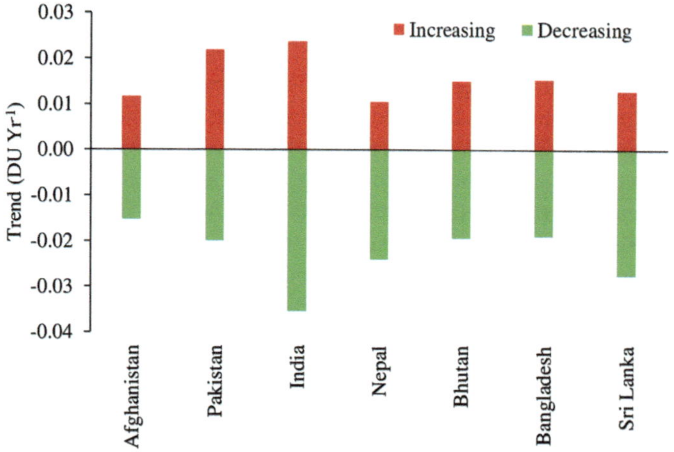

Fig. 9.8 Highest increasing and decreasing trend values of HCHO over South Asia, 2008–2018

data during 2008–2018. High HCHO was the observer over the Indo-Gangetic Plains (IGP) covering Bangladesh, India, and Pakistan. Concentration of HCHO was also found to be high in South Indian states viz., Andhra Pradesh, Kerala, and Telangana. The biogenic, thermal power plants, industries, pyrogenic (i.e. crop residue and biomass burnings), and fossil fuel combustions were likely culpable variables responsible for emission of HCHO in these areas (Baek et al. 2014; Biswas et al. 2019; Mahajan et al. 2015; Rana et al. 2019). However, post-harvest crop residue burning, was the primary reason for additional HCHO in the IGP region (Singh and Kaskaoutis 2014; Vadrevu et al. 2011). HCHO was found to be low during the

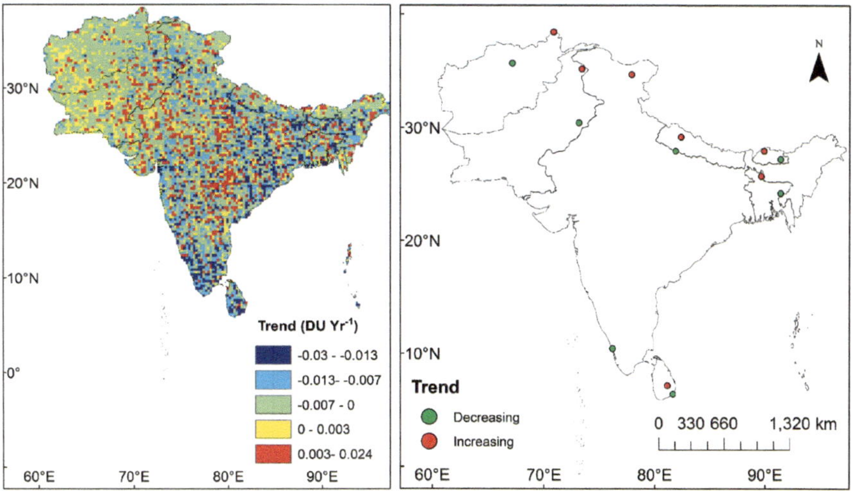

Fig. 9.9 The trend (left) and location of increasing and decreasing trend values (right) over South Asia, 2008–2018

monsoon season of July–September over South Asia due to rain-induced wash out process of trace gases from the atmosphere (David and Nair 2013; Lelieveld et al. 2018; Wang et al. 2018). HCHO tended to peak during June and November over South Asia. Pakistan, and Afghanistan that are the only countries among South Asia that peaked in July due to the late arrival of the monsoon (1st July) (Latif et al. 2016; Lu et al. 2018). HCHO follows a cyclic pattern; however, it does not follow the conventional seasonal cycle definition (Pielke et al. 1987). In addition, a decreasing trend was observed over all countries of South Asia in contrast to earlier studies during 1997–2009 (De Smedt et al. 2010). We found a significant decreasing trend over South Asia (-0.0641 DU yr^{-1}) but no specific year was detected as a change point. However, a statistically significant change point was detected for Bangladesh and Afghanistan in 2011.

Funding This study was funded by the Department of Science and Technology (DST), Government of India vide DST SERB No. EEQ/2021/000146 dated 02 March, 2022.

References

Abbot DS, Palmer PI, Martin RV, Chance KV, Jacob DJ, Guenther A (2003) Seasonal and interannual variability of North American isoprene emissions as determined by formaldehyde column measurements from space. Geophys Res Lett 30(17). https://doi.org/10.1029/2003gl017336

Alpert P, Osetinsky I, Ziv B, Shafir H (2004) A new seasons definition based on classified daily synoptic systems: an example for the eastern Mediterranean. Int J Climatol 24(8):1013–1021. https://doi.org/10.1002/joc.1037

Arlander DW, Brüning D, Schmidt U, Ehhalt DH (1995) The tropospheric distribution of formaldehyde during TROPZ II. J Atmos Chem 22:251–268

Azad AK, Kitada T (1998) Characteristics of the air pollution in the city of Dhaka, Bangladesh in winter. Atmosph Environ Dev Sustain 32(11):1991–2005

Badarinath KVS, Kiran Chand TR, Krishna Prasad V (2006) Agriculture crop residue burning in the Indo-Gangetic Plains-a study using IRS-P6 AWiFS satellite data. Curr Sci 91(8):1085–1089

Baek KH, Kim JH, Park RJ, Chance K, Kurosu TP (2014) Validation of OMI HCHO data and its analysis over Asia. Sci Total Environ 490:93–105. https://doi.org/10.1016/j.scitotenv.2014.04.108

Bhuvaneshwari S, Hettiarachchi H, Meegoda J (2019) Crop Residue burning in India: policy challenges and potential solutions. Int J Environ Res Public Health 16(5):832. https://doi.org/10.3390/ijerph16050832

Biswas MS, Ghude SD, Gurnale D, Prabhakaran T, Mahajan AS (2019) Simultaneous observations of nitrogen dioxide, formaldehyde and ozone in the indo-gangetic plain. Aerosol and Air Qual Res 19(8):1749–1764

Boeke NL, Marshall JD, Alvarez S, Chance KV, Fried A, Kurosu TP, … Millet DB (2011) Formaldehyde columns from the ozone monitoring instrument: urban versus background levels and evaluation using aircraft data and a global model. J Geophy Res 116(D5). https://doi.org/10.1029/2010jd014870

Bruni Zani N, Lonati G, Mead MI, Latif MT, Crippa P (2020) Long-term satellite-based estimates of air quality and premature mortality in Equatorial Asia through deep neural networks. Environ Res Lett 15(10):104088. https://doi.org/10.1088/1748-9326/abb733

Chan LY, Liu HY, Lam KS, Wang T (1998) Analysis of seasonal behaviour of tropospheric ozone at Hong Kong. Atmos Environ 1994(32):159–168

Chance K (2019) OMI/Aura Formaldehyde (HCHO) total column daily L3 weighted mean global 0.1deg Lat/Lon Grid V003. Retrieved from Greenbelt, MD, USA

Chaudhary QZ (1994) Pakistan summer monsoon rainfall's association with global and regional circulation features and its seasonal prediction. Paper presented at the international conference on monsoon variability and prediction, Triest, Italy

Chen X, Zhong B, Huang F, Wang X, Sarkar S, Jia S, … Shao M (2020) The role of natural factors in constraining long-term tropospheric ozone trends over Southern China. Atmos Environ 220:117060. https://doi.org/10.1016/j.atmosenv.2019.117060

Chhabra SK, Chhabra P, Rajpal S, Gupta RK (2001) Ambient air pollution and chronic respiratory morbidity in Delhi. Arch Environ Health Int Jo 56(1):58–64

Crutzen PJ, Andreae MO (1990) Biomass burning in the tropics: impact on atmospheric chemistry and biogeochemical cycles. Science 250(4988):1669–1678

Crutzen PJ, Heidt LE, Krasnec JP, Pollock WH, Seiler W (1979) Biomass burning as a source of atmospheric gases CO, H_2, N_2O, NO, CH_3Cl and COS. Nature 282(5736):253–256

Cusworth DH, Mickley LJ, Sulprizio MP, Liu T, Marlier ME, DeFries RS, … Gupta P (2018). Quantifying the influence of agricultural fires in northwest India on urban air pollution in Delhi, India. Environ Res Lett 13(4). https://doi.org/10.1088/1748-9326/aab303

David LM, Nair PR (2013) Tropospheric column O_3 and NO_2 over the Indian region observed by Ozone Monitoring Instrument (OMI): seasonal changes and long-term trends. Atmos Environ 1994(65):25–39. https://doi.org/10.1016/j.atmosenv.2012.09.033

De Smedt I, Müller JF, Stavrakou T, Eskes H, Roozendael MV (2008) Twelve years of global observations of formaldehyde in the troposphere using GOME and SCIAMACHY sensors. Atmos Chem Phys 8(16):4947–4963

De Smedt I, Stavrakou T, Müller JF, Van Der ARJ, Van Roozendael M (2010) Trend detection in satellite observations of formaldehyde tropospheric columns. Geophys Res Lett 37(18). https://doi.org/10.1029/2010gl044245

Field RD, van der Werf GR, Fanin T, Fetzer EJ, Fuller R, Jethva H, … Worden HM (2016). Indonesian fire activity and smoke pollution in 2015 show persistent nonlinear sensitivity to El Nino-induced drought. Proc Natl Acad Sci USA 113(33):9204–9209.https://doi.org/10.1073/pnas.1524888113

Fishman J, Wozniak AE, Creilson JK (2003) Global distribution of tropospheric ozone from satellite measurements using the empirically corrected tropospheric ozone residual technique: identification of the regional aspects of air pollution. Atmos Chem Phys 3:893–907
Fu T-M, Jacob DJ, Palmer PI, Chance K, Wang YX, Barletta B, … Pilling MJ (2007) Space-based formaldehyde measurements as constraints on volatile organic compound emissions in east and south Asia and implications for ozone. J Geophys Res 112(D6). https://doi.org/10.1029/2006jd007853
Gilbert RO (1987) Statistical methods for environmental pollution monitoring. Wiley, New York
Gurjar BR, Butler TM, Lawrence MG, Lelieveld J (2008) Evaluation of emissions and air quality in megacities. Atmos Environ 42(7):1593–1606
Helsel DR, Hirsch RM (1995) Statistical methods in water resources. Elsevier, New York
I PP, Jacob DJ, Chance K, Martin RV, Spurr RJ, Kurosu TP (2001) Air mass factor formulation for spectroscopic measurements from satellites: application to formaldehyde retrievals from the global ozone monitoring experiment. J Geophy Res 106(D13):14539–14550
Jaiswal RK, Lohani AK, Tiwari HL (2015) Statistical analysis for change detection and trend assessment in climatological parameters. Environ Process 2(4):729–749. https://doi.org/10.1007/s40710-015-0105-3
Kahya E, Kalaycı S (2004) Trend analysis of streamflow in Turkey. J Hydrol 289(1–4):128–144. https://doi.org/10.1016/j.jhydrol.2003.11.006
Kanaya Y et al (2009) Rates and regimes of photochemical ozone production over central East China in June 2006: a box model analysis using comprehensive measurements of ozone precursors. Atmos Chem Phys 9(20):7711–7723. https://doi.org/10.5194/acp-9-7711-2009
Krotkov NA, Lamsal LN, Celarier EA, Swartz WH, Marchenko SV, Bucsela EJ, … Wenig M (2017) The version 3 OMI NO2 standard product. Atmosph Measur Techniq 10(9). https://doi.org/10.5194/amt-10-3133-2017
Kumar V, Jain SK, Singh Y (2010) Analysis of long-term rainfall trends in India. Hydrol Sci J 55(4):484–496. https://doi.org/10.1080/02626667.2010.481373
Latif M, Syed FS, Hannachi A (2016) Rainfall trends in the South Asian summer monsoon and its related large-scale dynamics with focus over Pakistan. Clim Dyn 48(11–12):3565–3581. https://doi.org/10.1007/s00382-016-3284-3
Lelieveld J, Bourtsoukidis E, Bruhl C, Fischer H, Fuchs H, Harder H, … Neumaier M (2018) The South Asian monsoon-pollution pump and purifier. Sci Culture 361:270–273
Lieschke KJ, Fisher JA, Paton-Walsh C, Jones NB, Greenslade JW, Burden S, Griffith DWT (2019) Decreasing trend in formaldehyde detected from 20-year record at wollongong. Southeast Australia. Geophys Res Lett 46(14):8464–8473. https://doi.org/10.1029/2019gl083757
Lu X, Zhang L, Liu X, Gao M, Zhao Y, Shao J (2018) Lower tropospheric ozone over India and its linkage to the South Asian monsoon. Atmos Chem Phys 18(5):3101–3118. https://doi.org/10.5194/acp-18-3101-2018
MacDonald SM, Oetjen H, Mahajan AS, Whalley LK, Edwards PM, Heard DE, … Plane JMC (2012) DOAS measurements of formaldehyde and glyoxal above a south-east Asian tropical rainforest. Atmos Chem Phys 12(13):5949-5962.https://doi.org/10.5194/acp-12-5949-2012
Mahajan AS, De Smedt I, Biswas MS, Ghude S, Fadnavis S, Roy C, van Roozendael M (2015) Inter-annual variations in satellite observations of nitrogen dioxide and formaldehyde over India. Atmos Environ, 116.https://doi.org/10.1016/j.atmosenv.2015.06.004
Marlier ME, DeFries RS, Voulgarakis A, Kinney PL, Randerson JT, Shindell DT, … Faluvegi G (2013) El Nino and health risks from landscape fire emissions in Southeast Asia. Nat Clim Chang 3:131–136.https://doi.org/10.1038/nclimate1658
Martin RV, Fiore AM, Van Donkelaar A (2004) Space-based diagnosis of surface ozone sensitivity to anthropogenic emissions. Geophys Res Lett 31(6). https://doi.org/10.1029/2004gl019416
Mckendry IG, Steyn DG, O'Kane S, Zawar Reza P, Heuff D (1997) Lower tropospheric ozone measurements by light aircraft equipped with chemiluminescent sonde. J Atmosph Oceanic Technol 15:136–143

Meyer-Arnek J, Ladstatter-Weissenmayer A, Richter A, Wittrock F, Burrows JP (2005) A study of the trace gas columns of O_3, NO_2 and HCHO over Africa in September 1997. Faraday Discuss 130:387–405; discussion 491–517, 519–324. https://doi.org/10.1039/b502106p

Millet DB, Jacob DJ, Boersma KF, Fu T-M, Kurosu TP, Chance K, … Guenther A (2008) Spatial distribution of isoprene emissions from North America derived from formaldehyde column measurements by the OMI satellite sensor. J Geophy Res 113(D2). https://doi.org/10.1029/2007jd008950

Palmer PI (2003) Mapping isoprene emissions over North America using formaldehyde column observations from space. J Geophy Res 108(D6). https://doi.org/10.1029/2002jd002153

Pettitt AN (1979) A non-parametric approach to the change point problem. J R Stat Soc: Ser C: Appl Stat 28:126–135

Pettitt AN (1980) Some results on estimating a change-point using nonparametric type statistics. J Stat Comput Simul 11:261–272

Pielke RA, Garstang M, Lindsey C, Gusdorf J (1987) Use of a synoptic classification scheme to define seasons. Theoret Appl Climatol 38(2):57–68

Rana AD, Parvez S, Ul-Haq Z, Batool SA, Chaudhary MN, Mahmood K, Tariq S (2019) Anthropogenic, biogenic and pyrogenic emission sources and atmospheric formaldehyde (Hcho) and Nitrogen Dioxide (NO_2) columns over different landuse/landcovers of South Asia. Appl Ecol Environ Res 17(5). https://doi.org/10.15666/aeer/1705_1098911015

Salarijazi M (2012) Trend and change-point detection for the annual stream-flow series of the Karun River at the Ahvaz hydrometric station. African J Agric Res 7(32). https://doi.org/10.5897/ajar12.650

Salma S, Rehman S, Shah MA (2012) Rainfall trends in different climate zones of Pakistan. Pakistan J Meteorol 9(17)

Schott FA, McCreary JP Jr (2001) The monsoon circulation of the Indian Ocean. Prog Oceanogr 51(1):1–123

Sen PK (1968) Estimates of the regression coefficient based on Kendall's tau. J Am Stat Assoc 63:1379–1389

Singh RP, Kaskaoutis DG (2014) Crop residue burning: a threat to South Asian air quality. EOS Trans Am Geophys Union 95(37):333–334

Sinha B, Singh Sangwan K, Maurya Y, Kumar V, Sarkar C, Chandra BP, Sinha V (2015) Assessment of crop yield losses in Punjab and Haryana using 2 years of continuous in situ ozone measurements. Atmos Chem Phys 15:9555–9576. https://doi.org/10.5194/acp-15-9555-2015

Spinei E, Whitehill A, Fried A, Tiefengraber M, Knepp TN, Herndon S, … Poche B (2018) The first evaluation of formaldehyde column observations by improved Pandora spectrometers during the KORUS-AQ field study. Atmosph Measur Techniq 11(9):4943–4961.https://doi.org/10.5194/amt-11-4943-2018

Sullivan RC, Crippa P, Hallar AG, Clarisse L, Whitburn S, Van Damme M, … Pryor SC (2016) Using satellite-based measurements to explore spatiotemporal scales and variability of drivers of new particle formation. J Geophy Res Atmosph 121(20):12–217

Surl L, Palmer PI, González Abad G (2018) Which processes drive observed variations of HCHO columns over India? Atmos Chem Phys 18(7:4549–4566.https://doi.org/10.5194/acp-18-4549-2018

Theil H (1950) A rank-invariant method of linear and polynomial regression analysis, Part 1,2 & 3. Paper presented at the in proceedings of Koninalijke Nederlandse Akademie van Weinenschatpen A

Toro A, Richard A, Consuelo LO, Felipe M, Luis M, Raúl GE, Leiva G, Manuel A (2017) Trend and recovery of the total ozone column in South America and Antarctica. Clim Dyn 49(11–12):3735–3752. https://doi.org/10.1007/s00382-017-3540-1

ul-Haq Z, Tariq S, Ali M (2015) Tropospheric NO_2 trends over South Asia during the last decade (2004–2014) using OMI data. Adv Meteorol, 1–18. https://doi.org/10.1155/2015/959284

Vadrevu KP, Ellicott E, Badarinath KVS, Vermote E (2011) MODIS derived fire characteristics and aerosol optical depth variations during the agricultural residue burning season, north India. Environ Pollut 159(6):1560–1569

Van Belle G, Hughes JP (1984) Nonparametric tests for trend in water quality. Water Resour Res 20(1):127–136

Wang Y, Yu C, Tao J, Wang Z, Si Y, Cheng L, …Chen L (2018) Spatio-temporal characteristics of tropospheric ozone and its precursors in Guangxi, South China. Atmosphere 9(9):355. https://doi.org/10.3390/atmos9090355

Wang Y, Dörner S, Donner S, Böhnke S, De Smedt I, Dickerson RR, … Wagner T (2019). Vertical profiles of NO_2, SO2, HONO, HCHO, CHOCHO and aerosols derived from MAX-DOAS measurements at a rural site in the central western North China Plain and their relation to emission sources and effects of regional transport. Atmos Chem Phys 19(8):5417–5449. https://doi.org/10.5194/acp-19-5417-2019

Wijngaard JB, Klein Tank M, Konnen GP (2003) Homogeneity of 20th century European daily temperature and precipitation series. Int J Climatololgy 23:679–692

Yarragunta Y, Srivastava Sh, Mitra D, Le Flochmoën E, Barret B, Kumar P, Chandola HC (2019) Source attribution of carbon monoxide and ozone over the Indian subcontinent using MOZART-4 chemistry transport model. Atmosph Res 227:165–177. https://doi.org/10.1016/j.atmosres.2019.04.019

Zhang YH et al (2008) Regional ozone pollution and observation-based approach for analyzing ozone-precursor relationship during the PRIDE-PRD2004 campaign. Atmos Environ 42:6203–6218. https://doi.org/10.1016/j.atmosenv.2008.05.002

Chapter 10
Vegetation Sensitivity to Changing Climate in Bangladesh Using SPOT-VGT NDVI Time Series Data

Trishna Changkakati

Abstract The Normalized Difference Vegetation Index (NDVI) has found a wide application in vegetative studies as it has been used to estimate changes in vegetation health, vegetation cover, croplands, drought, grasslands, rangeland, etc. It is also associated with surface water, biomass range, photosynthetic level, and leaf area index. The main objective of this study is to assess the variability of vegetation health and its relationship with climatic attributes in Bangladesh. Here, 10-day SPOT-Vegetation Normalized Difference Vegetation Index (NDVI) time series data for the period 1998 to 2013 with 1 km resolution was used to study the vegetation changes in Bangladesh on the basis of 26 study sites based on the GLC 2009 land use categories. The frame work for the analysis is the use of correlation and coefficient of determination (R2) to determine the temporal changes of NDVI based on 800 spatially distributed random points in Bangladesh for the same period. A Pearson's correlation coefficient was used to define the relationship of monthly growing season NDVI with monthly rainfall and temperature derived from CRU TS 3.2 datasets. And analysis of 30- and 60-day lag rainfall was also figured. The result shows that there was a fluctuation in the vegetation growth during the wettest months for each year. The analysis between climatic attributes and NDVI during the period 1998–2013 showed a negative correlation. Thus from the analysis it emerged that the negative correlation between NDVI and precipitation and slightly positive correlation with temperature was emphasized during the growing season.

Keywords Vegetation dynamics · Rainfall · Temperature · SPOT-vgt · Bangladesh

10.1 Introduction

NDVI is generally accredited as a good indicator of terrestrial vegetation productivity. It is used to measure the amount of above ground green vegetation cover from measurements of red and near-infrared reflectance. It is extensively used by

T. Changkakati (✉)
Department of Earth Science, University of Science and Technology, Meghalaya, India
e-mail: trish.cgk@gmail.com

A. Saikia and P. Thapa (eds.), *Environmental Change in South Asia*,
https://doi.org/10.1007/978-3-030-47660-1_10

researchers to estimate above ground green biomass and also specifies the level of photosynthetic active radiation of vegetation cover. The SPOT-Vegetation Normalized Difference Vegetation Index (NDVI) has been widely used as an effective indicator of plant dynamics in studying the relationships between climatic elements and vegetation activity. The SPOT-vegetation sensor carried aboard SPOT 4 and 5 with 1 km resolution was designed to observe the Earth on regional and continental scales. It is mostly essential for studying agriculture, deforestation, and other vegetation changes on a broad scale. In the vegetation indices, the near-infrared part of the spectrum reflects mostly in healthy vegetation because the internal mesophyll structure of green leaves strongly reflects NIR radiations and leaf chlorophyll and other pigments absorb a large proportion of the red VIS radiation.

These vegetation indices are widely used and have benefited several disciplines concerned in assessment of biomass, water availability, and plant health and crop production (Jackson and Huete 1991). NDVI is usually recognized as a virtuous indicator in worldwide for vegetation productivity. Understanding effect in factors of climate, especially in rainfall and temperature on NDVI allows forecast of productivity changes under different climatic scenarios (Wang et al. 2010). NDVI provides an estimation of vegetation health and a way of detecting changes in vegetation cover over long period of time. The range of NDVI is between −0.1 (NIR less than VIS for a not very green area) and 0.6 (for a very green area).

The research work presented here is intended to understand the green biomass effect and its relationships with two important climatic attributes, namely, rainfall and temperature in Bangladesh for period of 1998 to 2013. As a result, the interannual, seasonal, and spatial variations of NDVI and their relationships with climate (temperature and rainfall) in Bangladesh were observed by using various satellite imageries and GIS techniques. The rationale for preferring to use temperature and rainfall lies in the fact that these two are the prime climatic attributes that influence the vegetation vigor and health in an area (Thomte et al. 2022; Príncipe et al. 2022; Sharma et al. 2022). While certain studies also tend to incorporate soil moisture as an additional variable that influences NDVI, this analysis was confined to the influence of temperature and precipitation on NDVI.

10.2 Study Area

The People's Republic of Bangladesh is located in South Asia. The country shares its land boundary with India to the west, north, and east; Myanmar to the southeast; and the Bay of Bengal to the south. It extends from 20°34′N to 26°38′N latitude and 8°01′E to 92°41′E longitude and covers an area of 143,998 km^2. The country is conquered by the fertile ranges of Ganges–Brahmaputra Delta, which is the largest river delta in the world. The northwest and central part of the country comprises the Madhupur and therefore the Barind Plateaus. The northeastern and southeastern regions of Bangladesh are abode to evergreen hill ranges. Around 17% of the country is covered by forests and 12% is covered by hill systems (Jackson and Huete 1991).

The country's climatic condition is especially of warm and humid monsoon type which persists from June to October and supplies about 80% of its rainfall. The annual rainfall varies from maximum of 5690 mm within the northeastern part and minimum of 1110 mm within the western part of the country. On the other hand, the average temperature is about 25 °C. The maximum temperature in summer varies between 38 °C and 41 °C April is the hottest month of the country (Monirul et al. 2015). The eastern part of the country comprised of some parts of Sylhet and Chittagong occupies small hills and covered by forest. The north of Dhaka supports swampy vegetation in the central part of the country. The northwest and southwest zones comprised mostly of cultivated plants and orchards whereas the southern zone along the Bay of Bengal comprises wetlands with mangrove vegetation (Monirul et al. 2015). Large areas of forest lie within the Chittagong Hills and the Sundarbans. The evergreen and deciduous forest of the Chittagong Hills cover a vast amount of teak used for heavy construction, boat building, etc. Sundarbans, a tidal mangrove forest covering nearly 6,000 sq km along the Bay of Bengal, is the source of timber used for various purposes in the country's economy (Fig. 10.1).

10.3 Methods and Analysis

In this study, satellite imageries of NDVI time series derived from the Systeme Pour l' Obsevation de la Tere (SPOT) Vegetation S-10 NDVI data with 1 km resolution for the period 1998–2013 were used to analyze vegetation health in Bangladesh. S10 dataset is 10 days synthesized data derived from S1 products based on the NDVI Maximum Value Composite (MVC) algorithm, not taking into account the bad quality data or cloud masked data (VEGETATION users guide 1999). It is predominantly essential for studying agriculture, deforestation, and other vegetation changes on a broad scale. SPOT Vegetation products are systematically acquired, archived, and can be downloaded easily from its website. All the satellite imageries were georeferenced and geometrically corrected in ArcGIS 10.1 software and were thereafter used for further analysis.

10.3.1 Time Series Datasets

NDVI is generally accepted as a good indicator of terrestrial vegetation productivity. The influence of climatic factors, in particular, rainfall and temperature on NDVI allows prediction of productivity changes under different climatic scenarios (Wang et al. 2010). NDVI is calculated as the difference between near-infrared (NIR) and visible (VIS) reflectance values normalized over the sum of the two.

$$\mathrm{NDVI} = (\mathrm{NIR} - \mathrm{VIS})/(\mathrm{NIR} + \mathrm{VIS})$$

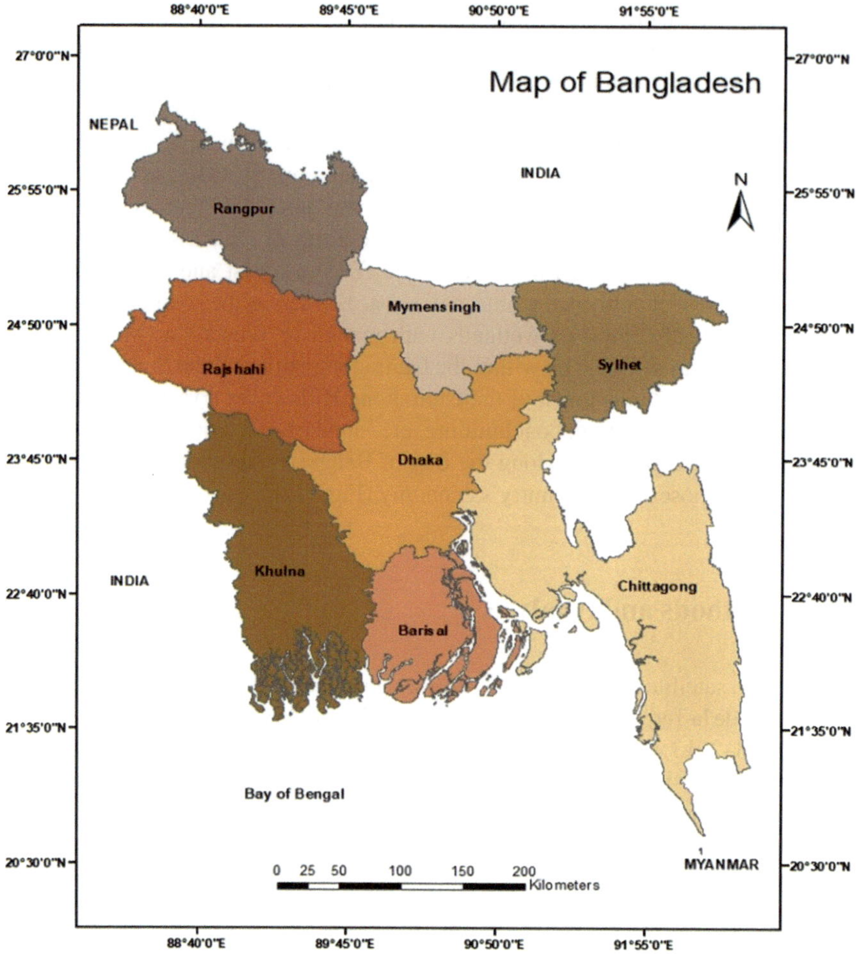

Fig. 10.1 Map of Bangladesh

The upper value of the NDVI approaches "1", while the lower value is generally close to "0". The lower value may be slightly positive or slightly negative, depending on sensor physiognomies and the units of the input variables such as radiance, reflectance, digital numbers, etc. This has long been used by various scholars to monitor the vegetation and changes in vegetation of the entire earth or parts of it thereof. NDVI for vegetation generally ranges from 0.3 to 0.8, with the larger values signifying "greener" surfaces and bare soils range from about 0.2–0.3 (Candiago et al. 2015; Bhandari et al. 2012, Xue and Su 2017). The NDVI time series data consist of total 576 raster images at 10-day intervals spanning 16 years, i.e., from 1998 to 2913. All the image pixel digital numbers (DN) were converted to ASCII and then they exported into spreadsheets to obtain actual value of NDVI using statistical equations to show its correlation with climate data. The monthly pixel values of NDVI were extracted by using random sampling method based on spatially distributed random points over the major land use categories provided by the regional Global Land Cover (GLC 2009) datasets (Balaghi et al. 2008; Camberlin et al. 2007), which was imported and analyzed in Arc GIS 10.1 software. However to observe the relationship between vegetation indices and rainfall and temperature, the (Climatic Research Unit-time series) CRU TS 3.2 datasets based on 800 spatially distributed random points for the period 1998–2013 were used. The station records of rainfall and temperature were obtained from randomly selected land use and landcover categories based on GLC 2009 dataset.

10.3.2 *Analysis of the Correlation Between NDVI, Rainfall, and Temperature*

Pearson's product-moment correlation coefficient (or Pearson's correlation coefficient, for short) is a measure of the strength of a linear association between two variables and is denoted by r (Pearson 1895). Basically, Pearson's product-moment correlation attempts to draw a line of best fit through the data of two variables, and Pearson's correlation coefficient, r, indicates how far away all these data points are to this line of best fit. It is given by the formula

$$\boldsymbol{r} = \frac{-(\boldsymbol{x}_i - \overline{\boldsymbol{x}})(\boldsymbol{y}_i - \overline{\boldsymbol{y}})}{\boldsymbol{n}.\sigma \boldsymbol{x} \sigma \boldsymbol{y}} \tag{10.1}$$

where

r = coefficient of correlation,
x_i= values of the variables in one distribution,
$\overline{x}$ = mean of the distribution,
y_i= values of the variables of other distribution,
$\overline{y}$= mean of the above distribution,
n = number of observation,
σx = standard deviation of x-distribution, and
σy = standard deviation of y-distribution.

Here, xi refers to monthly rainfall or temperature data and yi refers to monthly NDVI data, $\overline{x}$ refers to average of the rainfall data and $\overline{y}$ refers to average of NDVI, and σx and σy is the standard deviation of rainfall and NDVI. All the components were derived using a spreadsheet and thereafter Pearson's correlation coefficient was calculated and plotted in a graph. Pearson's product–moment linear correlation (r values having a range from −1 to 1) is used for testing trends by estimating the linear correlation between the values of each pixel over time and a perfectly linear series (Fensholt and Proud 2012). A value of 0 indicates that there is no association between the two variables. A value greater than 0 indicates a positive association, that is, as the value of one variable increases, so does the value of the other variable

(Fensholt and Proud 2012). A value less than 0 indicates a negative association, that is, as the value of one variable increases, the value of the other variable decreases (Pearson 1895). Correlation coefficients between NDVI and rainfall of year-to-year time duration and 30 days and 60 days lag were also computed in different land use categories of Bangladesh with the help of spreadsheets.

10.4 Results and Discussions

10.4.1 Interannual Variations in NDVI Variability

The monthly and annual configurations in NDVI variability over Bangladesh for the period 1998–2013 were computed on the basis of 800 randomly selected points that are spatially distributed all over the country.

In this graph, each NDVI curve for the 16-year period is represented by its time series. The trend (Fig. 10.2) expresses a long time of expanded and diminished vegetation growth in the entire region. The NDVI value extends between 0.55 and 0.67 in the months of September and October which is the peak month for vegetation growth in the years 2003, 2006, 2009, 2010, 2011, and 2013. The rest of the years contain less than 0.55 units of NDVI. In the wettest month, i.e., June and July the value of vegetation indices was less than 0.40 units which reveals the effect of rainfall variability on vegetation growth. These months exhibited variations in vegetation vigor and greenness during the growing season in the study area which might be due to ecological effects beyond those brought about by precipitation.

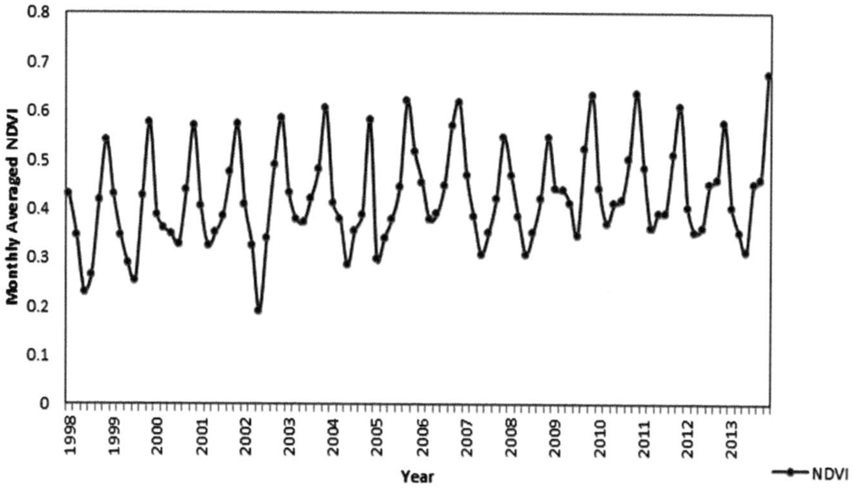

Fig. 10.2 Monthly NDVI time series averaged over the entire Bangladesh, 1998–2013

10.4.2 Correlation Between NDVI and Climatic Attributes (Rainfall and Temperature)

To analyze the spatial pattern of NDVI and their relationship with climate, maps of mean NDVI and growing season NDVI were compared with the interpolated rainfall and temperature maps (Fig. 10.4).

This interpolation was derived with the help of 800 randomly selected points based on GLC 2009 image and was computed in Arc GIS 10.2.1 software (Fig. 10.3).

The highest rainfall was observed in Khulna, Chittagong, Dhaka, central part of Rajshahi, Rangapur, few parts of Mymensingh, and a Sylhet region. Most parts of Bangladesh are covered with irrigated croplands that seem to be an agricultural land, shrublands and mosaic croplands were also an important land use of the country. The country is covered with a huge area of rainfed croplands and permanently flooded broadleaved forest in southeastern part.

The NDVI values in growing season lie between 0.73 (highest) and −0.06 (lowest), whereas in non-growing season the NDVI values range between 0.61 (highest) and −0.1 (lowest). Thus, it seems that the vegetation health during the growing season months, i.e., May–October is higher than that during the rest of the months in Bangladesh.

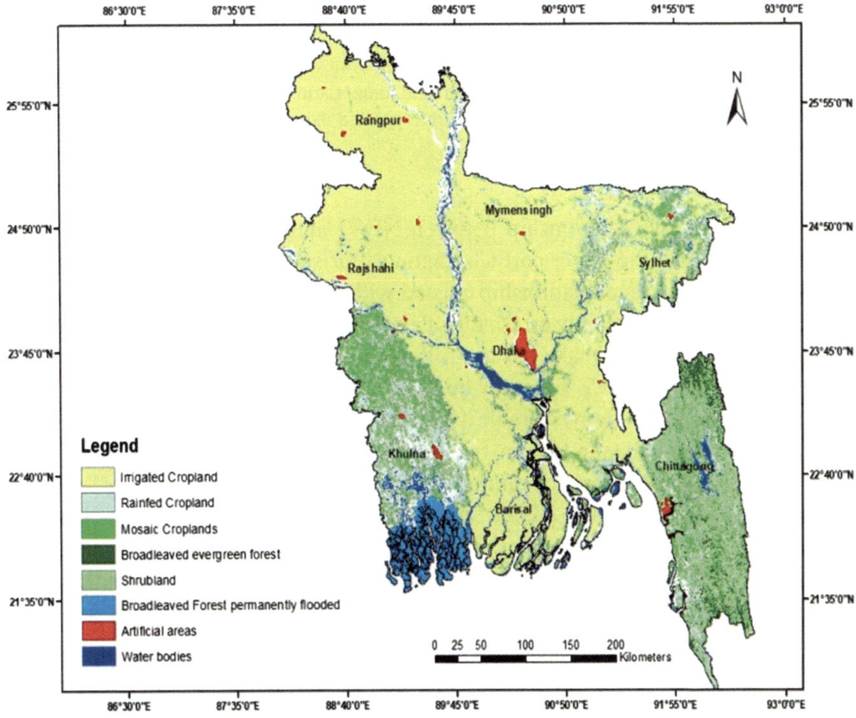

Fig. 10.3 Land use and landcover based on Globe Cover 2009 image

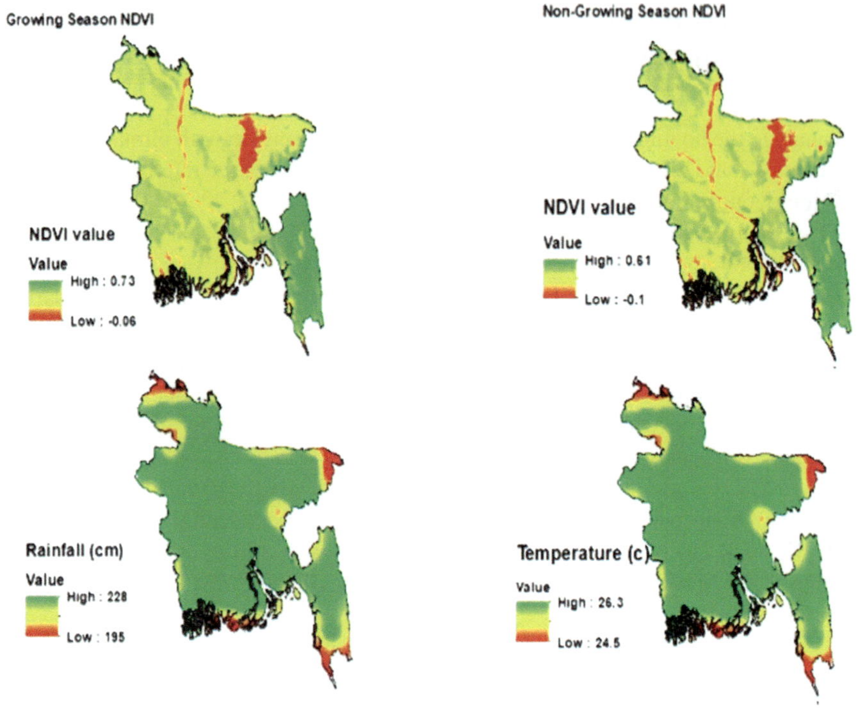

Fig. 10.4 Bangladesh: growing season NDVI, non-growing season NDVI, averaged Temperature and rainfall (clockwise) 1998–2013. The rainfall and temperature during the growing season (May, June, July, August, September, October) were interpolated using kriging, based on values of the 800 randomly selected points

There was inverse relationship between NDVI and climatic attributes (rainfall and temperature) in the lower part Chittagong, Barisal Sylhet, and upper region of Rangpur area. A positive relationship existed with areas where rainfall and temperature value was specifically lower. On the other hand, the central part of Mymensingh, Rajshahi, Rangpur, and Dhaka did not follow to the general inverse of relationship between rainfall and NDVI. This may be due to dominance of irrigated cropland and low percentage of forest cover (Table 10.1).

The above table indicates that the relationship between NDVI and rainfall is lower which means that the impact of rainfall is weaker. There was considerable year-to-year variation in rainfall and NDVI. The mean NDVI value in most years between 1998 and 2013 was almost similar. The highest NDVI value (0.476) was in 2006 and lowest NDVI value (0.371) was in 1998. Overall the NDVI values in the entire region lie between 0.674 and 0.229 which is not a decent indication for a healthy vegetation growth.

Here, linear correlation coefficients between NDVI and rainfall in Bangladesh were incorporated. NDVI was associated to the sum of monthly rainfall between1998

Table 10.1 Mean growing season rainfall and minimum and maximum NDVI and NDVI range in Bangladesh (1998–2013)

	NDVI mean	NDVI max	NDVI min	NDVI range	GSR mean (mm)
1998	0.371	0.54	0.229	0.31	203.17
1999	0.386	0.576	0.252	0.323	209.72
2000	0.405	0.569	0.326	0.243	209.66
2001	0.418	0.571	0.324	0.246	208.24
2002	0.389	0.586	0.189	0.396	210.78
2003	0.448	0.607	0.371	0.236	214.2
2004	0.399	0.58	0.284	0.296	209.91
2005	0.432	0.62	0.296	0.324	211.17
2006	0.476	0.619	0.376	0.242	215.81
2007	0.412	0.545	0.307	0.237	232.27
2008	0.412	0.545	0.307	0.237	230.46
2009	0.465	0.633	0.344	0.288	211.17
2010	0.462	0.637	0.369	0.267	209.66
2011	0.461	0.637	0.359	0.278	203.17
2012	0.441	0.674	0.311	0.363	210.78
2013	0.441	0.674	0.311	0.363	230.46

and 2013. With this process, the correlation coefficient trend of NDVI range values and growing season rainfall (GSR mean) from 1998 to 2013 was revealed in Fig. 10.5.

NDVI (Fig. 10.5). It is specified that NDVI considerably increased at a rate of 0.0009 over GSR mean ($R^2 = 0.0238$, $p = 0.0643$) from 1998 to 2013. This P value of

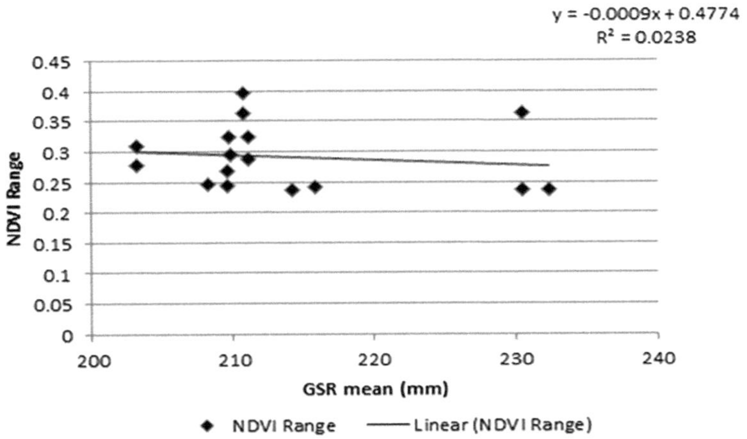

Fig. 10.5 Linear relationships between mean growing season rainfall and NDVI range

0.64 being more than the benchmark of 0.05 and low R^2 combination indicates non-significance of the test. The NDVI range value is from 0.236 to 0.396 and the GSR mean value is from 203.17(mm) to 232.27 (mm). The highest rainfall, i.e., 232.27 mm occurred in 2007 and its NDVI value is 0.237 and the lowest rainfall is 203.17 mm occurred in 1998 and its NDVI value is 0.31. Therefore, it is specified that response of rainfall to NDVI shows negative correlation. It is mentioned that variations in NDVI values among the curves may reveal the effect of rainfall variability on vegetation growth, which may be significance of climate changes (Changkakati 2018).

According to Luo et al. (2009), there were strong correlations between NDVI, rainfall, and temperature for different vegetation forms in northeast China, where the influence of temperature on NDVI was more obvious than that of rainfall. Therefore, to analyze the correlation between temperature and NDVI, a linear regression process was opted for in the current analysis.

Here, linear correlation coefficients between NDVI and temperature in eight land use categories of 26 sites in Bangladesh are shown. The linear correlation coefficient between NDVI and temperature was examined. NDVI was strongly related to the sum of monthly temperatures between 1998 and 2013. Following this process, the correlation coefficient trend of NDVI range values and mean temperature of the growing season from 1998 to 2013 shown in Fig. 10.6 were examined. The averaged growing season NDVI and monthly temperature when plotted in a scatter diagram showed a negative linear relationship. While linear relationship was explored it showed R2 value of 0.0822 which indicates positive correlation (see Fig. 10.6). It indicates that NDVI is significantly increasing at a rate of 0.0132/year where (R2 = 0.0822, P = 0.0422). This P value of 0.0.422 being less than the benchmark of 0.05 and R^2 combination indicates the significance of the test. Following previous studies the effects of rainfall and temperature on NDVI varied for different vegetation types (Chuai et al. 2013). High temperature tends to be linked with a healthy vegetation growth and a positive relation between NDVI and temperature exists (Fig. 10.6).

10.5 Within Season Relationship Between NDVI and Rainfall

There was a substantial yearly variation in rainfall and NDVI. The linear regression result shows negative correlation between NDVI range and averaged growing season rainfall for the period 1998–2013. To explore stronger and more significant relationships, 30-day lag and 60-day lag rainfall data was used and its relation to averaged monthly NDVI for the same period was examined. Previous studies on the response of NDVI in comparatively lower rainfall regimes establish good correlation between NDVI and rainfall lags specially in drought regions (Saikia 2009; Keriyeva et al. 2011; Piao et al. 2006; Barbosa et al. 2006; Barbosa et al. 2016) and such a lag was marked in Bangladesh, although the relationship was negative possibly on account of the rather high rainfall received (Table 10.2).

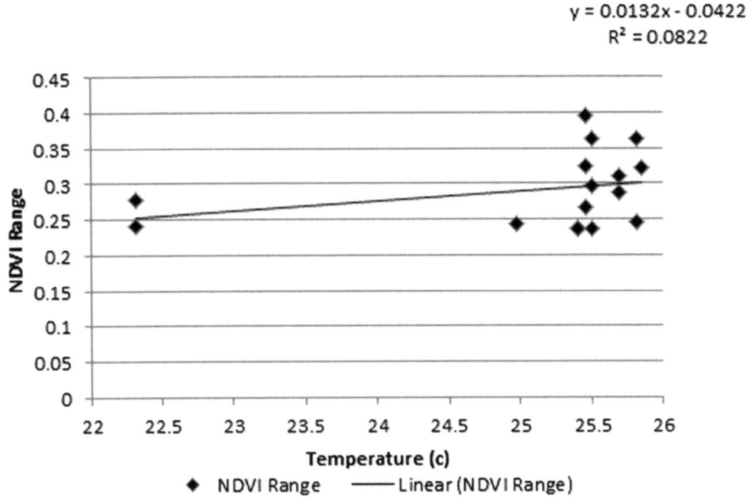

Fig. 10.6 Linear relationships between averaged monthly temperature and monthly NDVI range

When the monthly rainfall versus monthly NDVI was calculated, correlation became more prominent as 21 sites out of 26 sites showed significant correlations. On using 1-month rainfall lag, 15 sites out of 26 sites showed significant correlations. A 2-month lag rainfall showed that 20 of 26 sites possessed a significant correlation. Mainly irrigated cropland, mosaic cropland, rainfed cropland, broadleaved evergreen forest, and broadleaved permanently flooded forest revealed significant correlations. The response of rainfall to diverse vegetation classes is not distinct between land cover classes. However, in the monthly and 2 monthly lag, a positive correlation was evident in a few sites. Whereas the 1-month lag correlation coefficient analysis for all the land use categories revealed slight improvements, and a greater effect was seen with a 2-month lag.

10.6 Conclusion

The overall analysis of trend in vegetation greenness and vegetation health for Bangladesh was observed where the NDVI value extends between 0.55 and 0.67 in the months of September and October which is the peak period for vegetation growth in the years 2003, 2006, 2009, 2010, 2011, and 2013. The rest of the year exhibits less than 0.55 units of NDVI. In the wettest month, i.e., June and July, the value of vegetation indices was less than 0.40 units which reveals the effect of rainfall variability on vegetation growth. The research indicates fluctuation in NDVI value in the growing season based on monthly and annual variations in the entire region of

Table 10.2 Coefficient of determination (R^2) for the correlations between NDVI and rainfall

	Land use/ land cover	1	2	3
1	Irrigated croplands	0.431**	0.371**	0.463**
2	Irrigated croplands	0.513	0.015**	0.24
3	Irrigated croplands	0.435	0.36	0.273**
4	Irrigated croplands	0.418**	0.344**	0.343**
5	Irrigated croplands	0.359**	0.406	0.257**
6	Irrigated croplands	0.433**	0.319	0.323**
7	Shrublands	0.457	0.322	0.327**
8	Irrigated croplands	0.398**	0.275	0.279
9	Mosaic croplands	0.395**	0.309	0.251**
10	Irrigated croplands	0.424**	0.312	0.291
11	Mosaic croplands	0.551**	0.389**	0.378**
12	Irrigated croplands	0.450**	0.257	0.239
13	Irrigated croplands	0.368**	0.334**	0.391**
14	Irrigated croplands	0.456**	0.341**	0.398**
15	Irrigated croplands	0.474**	0.302**	0.401**
16	Mosaic croplands	0.330**	0.371	0.171**
17	Artificial areas	0.409**	0.329**	0.331**
18	Irrigated croplands	0.444**	0.301	0.289
19	Rainfed croplands	0.345**	0.264**	0.399
20	Mosaic croplands	0.398**	0.353	0.317**
21	Broadleaved evergreen forest	0.474**	0.331**	0.376**
22	Rainfed croplands	0.346	0.342**	0.374**
23	Broadleaved forest permanently flooded	0.452	0.335**	0.373**
24	Irrigated croplands	0.402**	0.402**	0.439**
25	Mosaic croplands	0.430**	0.440**	0.338**
26	Irrigated croplands	0.378**	0.409**	0.429**

Note The header columns numbered 1–3 refer to correlations between
1. Monthly rainfall versus monthly NDVI
2. 1-month lag averaged monthly rainfall with averaged monthly NDVI
3. 2-month lag averaged monthly rainfall with averaged monthly NDVI
When the monthly rainfall versus monthly NDVI was calculated, correlation became

Bangladesh. The relationship between variations in remote sensing indices and environmental variables allows for an effective estimation of the main environmental impacts, which can be used for coastal planning and management (Alatorre et al. 2016).

While temperature and precipitation are important determinants of vegetation greenness and health employed in this analysis, often other drivers are also at play: these are soil processes and local hydrology (Crichton et al. 2022). However, these agents (soil processes and local hydrology) were beyond the scope of this analysis. Future analysis related to NDVI in Bangladesh could certainly incorporate these drivers. Additionally, crowdsourcing and Google Earth have been used to map diverse corners of the Earth's wilderness areas (See et al. 2016) and these methods can be fruitfully employed in Bangladesh and other countries as well. In fact such methodologies also enable the visualization of spatial human impact extents (See et al. 2016) and thus serve a dual function. Similarly, open-source datasets like Landsat and Sentinel have enabled the mapping of urban built landscapes as well (See et al. 2022) and assessing urban NDVI trends is now more easily possible than ever before.

Our analysis refers to September–October as months with peak vegetation growth, whereas May to October is the growing season period. This is quite common in literature related to NDVI and vegetation growth. Generally growing and no-growing seasons are delineated on the basis of the precipitation regime, whereas the peak vegetation growth is evidenced by the months with the highest NDVI values.

The comparisons between Normalized Difference Vegetation Index (NDVI) with 1 km spatial resolution are suitable for monitoring vegetation condition in areas where rainfall and temperature are high. The mean monthly temperature shows a considerable impact on NDVI variation. However, rainfall has fewer effects on NDVI than temperature with a lag of 30 and 60 days (Li et al. 2012). This could be explained by the fact Bangladesh receives generous amounts of precipitation and thus temperature and its variability is a determining factor. The condition of Bangladesh is not dissimilar to that of northeast India. The climatic attributes, i.e., precipitation and temperature variables showed the largest impact among other environmental variables on vegetation dynamics across the region, and projected climate changes could rigorously impact water use and agricultural production (Keriyeva et al. 2011). According to the Ministry of Environment and Forest (MoEF) of Bangladesh, 11.1% or about 1,442,000 ha of Bangladesh is covered by forest. Of this 30.2% (436,000) is classified as primary forest, the most biodiversity and carbon-dense form of forest. But in the present era the forest areas have been deforested by anthropogenic drivers. One of the primary causes of deforestation and forest fragmentation in Bangladesh is the increasing rate of population growth. In addition, the demand of forest products and fuel wood has been tremendously increasing which stokes deforestation. Conversion of forest land to agricultural land, industrial, urbanization, and development of infrastructures for transportation, energy, etc. also results in the loss of forest cover. This analysis brought out the changes to ecological balance through NDVI-based change detection in large tracts of Bangladesh and assessed loss in forest quality by using remotely sensed indicators of biomass productivity. Such degradation to Bangladesh's forests does not auger well for the flora and fauna, as studies in neighboring India (Ahmed et al. 2022) shows that land use change and habitat fragmentation deleteriously affect flagship species. With climate change adding the Earth's woes, the time is nigh for Bangladesh to give increased attention to her forest health.

References

Ahmed R, Saikia A, Robeson SM (2022) Tracks of death: elephant casualties along the Habaipur–Diphu railway in Assam, India. Ann Amer Assoc of Geogr 112. https://doi.org/10.1080/24694452.2021.1990009

Alatorre LC, Carrillo SS, Beltran SM, Medina RJ (2016) Temporal changes of NDVI for qualitative environmental assessment of mangroves: Shrimp farming impact on the health decline of the arid mangroves in the Gulf of California (1990–2010). J Arid Environ 125:98–109. https://doi.org/10.1016/j.jaridenv.2015.10.010

Balaghi R, Tychon B, Eerens H, Jlibene M (2008) Empirical regression models using NDVI rainfall and temperature data for the early prediction of wheat grain yields in Morocco. Int J Appl Earth Obs Geoinf 10:438–452

Barbosa HA, Huete AR, Baethgen WE (2006) A 20-year study of NDVI variability over the Northeast region of Brazil. J Arid Environ 67:288–307

Barbosa HA, Kumar, Lakshmi TV (2016) Influence of rainfall variability on the vegetation dynamic over Northeastern Brazil. J Arid Environ 124:377–387

Bhandari AK, Kumar A, Singh GK (2012) Feature extraction using normalized difference VegetationIndex (NDVI): a case study of Jabalpur City. In: 2nd international conference on communication, computing & security [ICCCS-2012], pp 612–621

Camberlin P, Martiny N, Philippon N, Richard Y (2007) Determinants of the interannual relationships between remote sensed photosynthetic activity and rainfall in tropical Africa. Remote Sens Environ 106:199–216

Candiago S, Remondino F, Gigilo MD, Dubbini M (2015) Evaluating multispectral images and vegetation indices for precision farming applications from UAV images. Remote Sens 7(4):4026–4047. https://doi.org/10.3390/rs70404026

Changkakati T (2018) Spatial and temporal patterns of NDVI variability in North East (1998–2013). J Gujarat Res Soc 21(1):74–82. ISSN: 0374–8588

Chuai XW, Huang XJ, Wang WJ, Bao G (2013) NDVI, temperature and precipitation changes and their relationships with different vegetation types during 1998–2007 in Inner Mongolia, China. Int J Climatol 33:1696–1706

Fensholt R, Proud SR (2012) Evaluation of earth observation based global long term vegetation trends—comparing GIMMS and MODIS global NDVI time series. Remote Sens Environ 119:131–147

Islam Monirul Md, Mamum I, Mainul Md (2015) Variations of NDVI and its association with rainfall and evapotranspiration over Bangladesh. Rajshahi Univ J Sci Eng 43:21–28. ISSN 2309–0952

Jackson RD, Huete AR (1991) Interpreting vegetation indices. Prev Vet Med 11(3–4):185–200

Kariyeva J, Leeuwen, Willem JD (2011) Environmental drivers of NDVI—based vegetation phenology in central Asia. Remote Sens 3:203–246. https://doi.org/10.3390/rs3020203

Li Z, Guo X (2012) Detecting climate effects on vegetation in northern mixed prairie using NOAA AVHRR 1-km time-series NDVI data. Remote Sens 4:120–134. https://doi.org/10.3390/rs4010120

Luo L, Wang ZM, Song KS, Zhang B, Liu DW, Ren CY, Zhang SM (2009) Research on the correlation between NDVI and climatic factors of different vegetation in the northeast China. Xibei Zhiwu Xuebao 29(4):800–808

Pearson K (1895) Notes on regression and inheritance in the case of two parents. Proc Royal Soc Lond 58:240–242

Piao S, Mohammat A, Fanga J, Caia Q, Feng J (2006) NDVI-based increase in growth of temperate grasslands and its responses to climate changes in China. Glob Environ Chang 16:340–348

Príncipe A, Nunes A, Pinho P, Aleixo C, Neves N, Branquinho C (2022) Local-scale factors matter for tree cover modelling in Mediterranean drylands. Sci Total Environ 831:154877

Saikia A (2009) NDVI Variability in North East India. Scottish Geogr J 125:195–213

See L, Fritz S, Perger C, Changkakati T, Obersteiner M (2016) Mapping human impact using crowdsourcing. In: Carver SJ, Fritz S (eds) Mapping wilderness: concepts, techniques and applications. Springer, Dordrecht, pp 89–101

See L, Georgieva I, Duerauer M, Karner M, Fritz S (2022) A crowdsourced global data set for validating built-up surface layers. Scientific Data 9(1):13

Sharma M, Bangotra P, Gautam AS, Gautam S (2022) Sensitivity of normalized difference vegetation index (NDVI) to land surface temperature, soil moisture and precipitation over district Gautam Buddh Nagar, UP, India. Stoch Env Res Risk Assess 36(6):1779–1789

Thomte L, Bhagabati AK, Shah SK (2022) Soil moisture-based winter–spring drought variability over West Karbi Anglong region, Assam, Northeast India using tree-rings of Pinus kesiya. Environ Challenges 7:100512

Wang J, Meng JJ, Cai YL (2008) (2010) Assessing vegetation dynamics impacted by climate change in the South western karst region of China with AVHRR NDVI and AVHRR NPP time-series. Environ Geol 54:1185–1195

Xue J, Su B (2017) Significant remote sensing vegetation indices: a review of developments and applications. J Sens, Article ID 1353691, 17. https://doi.org/10.1155/2017/135369

Chapter 11
When the Bough Breaks: Spatial Variability of Tropospheric Ozone in the Indian Sub-continent

Anup Saikia, Rekib Ahmed, Kiran Sharma, Khangsembo Bungnamei, Chandra Kant Pawe, Sourav Chetia, Ujjal D. Baruah, Prasanta Bhattacharyya, Syeda Rizwana Yasmin, Nitashree Mili, Dhrubajyoti Sahariah, and Pankaj Thapa

Abstract Tropospheric or ground level ozone (O_3) is an important greenhouse gas and a pollutant harmful to human health and agricultural activity. With high population pressure in the Indian Sub-continent and economies heavily reliant on agriculture and allied activities tropospheric ozone is problematic. This analysis assesses the spatial variability of tropospheric ozone using monthly AIRX3STM satellite data during 2002–16. The monthly images were combined and averaged to derive month-wise and four season-wise aggregates during 2002–2016 in order to understand monthly and seasonal variations in the spatial patterns of O_3. A gain coefficient image was also generated using a GIS to visualize and depict areas with spatial variability of O_3 during the period.

Keyword Tropospheric ozone · AIRX3STM satellite data · Indian Sub-continent

---

A. Saikia (✉) · R. Ahmed · K. Sharma · K. Bungnamei · S. Chetia · U. D. Baruah · P. Bhattacharyya · S. R. Yasmin · D. Sahariah
Department of Geography, Gauhati University, Guwahati, India
e-mail: asaikia@gauhati.ac.in

K. Bungnamei
e-mail: ksmbmei@gauhati.ac.in

S. Chetia
e-mail: schetia@gauhati.ac.in

U. D. Baruah
e-mail: udbmail@gauhati.ac.in

D. Sahariah
e-mail: dhrubajyoti@gauhati.ac.in

C. K. Pawe
Department of Geography, Pragjyotish College, Guwahati, India

U. D. Baruah · N. Mili
Department of Geography, Cotton University, Guwahati, India

P. Thapa
Sherubtse College, Royal University of Bhutan, Kanglung, Bhutan

A. Saikia and P. Thapa (eds.), *Environmental Change in South Asia*,
https://doi.org/10.1007/978-3-030-47660-1_11

11.1 Introduction

Tropospheric or ground level ozone (O_3) is an important greenhouse gas and a pollutant harmful to human health and agricultural activity (Anderson et al. 2016). Tropospheric ozone is not added to air directly, but forms photochemically involving precursors such as carbon monoxide (CO), volatile organic compounds (VOCs), and oxides of nitrogen (NOx), supplemented by transport from the stratosphere (Sharma et al. 2017; Bhuyan et al. 2014; Ghude et al. 2014; Sitch et al. 2007). The adverse effects of ozone are particularly significant near the Earth's surface (Sharma et al. 2017) and elevated surface ozone is detrimental to human health and ecosystem health (Bourgeois et al. 2021) and adversely affects and causes premature mortality (Ebi and McGregor 2008; Jerrett 2015; Lelieveld et al. 2015), aggravating pulmonary and respiratory diseases (Ebi and McGregor 2008).

The thickness of the ozone layer is measured in Dobson units (DU), which assesses what the physical thickness would amount to when compressed under atmospheric conditions within the Earth's atmosphere (Hanslmeier 2008). G.M.B. Dobson was one of the first scientists to examine atmospheric ozone (Roaf et al. 2009) and developed an instrument to measure ozone a century ago. The Dobson spectrophotometer measures the intensity of sunlight at two ultraviolet wavelengths and one DU is defined to be 0.01 mm thickness under standard conditions of temperature and pressure (Rajab et al. 2013).

Emissions from burning fossil fuels as well as biomass have caused elevated ambient ozone concentrations over large pockets across the Earth's surface (Sitch et al. 2007).

The role of biomass burning in controlling tropical O_3 is widely established (Anderson et al. 2016; Sauvage et al. 2007; Wang et al. 2001) and greater tropospheric ozone over the Southern Atlantic during winter has been associated with biomass burning in the northern hemisphere over Africa (Monks et al. 2009). Likewise, biomass burning in southern Africa has been linked to enhanced ozone levels in the northern hemisphere over West Africa in austral winter. Biomass burning in Indonesia has been known to affect the boundary layer over northern Australia with elevated concentrations of NOx and ozone aloft, with pollutants being transported from Indonesia (Monks et al. 2009; Kondo et al. 2004). Thus, accumulation of O_3 at distances up to thousands of kilometers downwind from the initial polluting source resulting in enhanced concentrations even in areas located remote from the sources of precursors is not uncommon (Emberson and Büker 2008; Sharma et al. 2017). Winds transport O_3 to areas thousands of kilometers away and areas without significant precursor emissions are often adversely impacted (Engardt 2008).

This makes O_3 a pollutant more likely to affect agricultural regions with elevated O_3 concentrations; a pollutant that is likely to threaten global food production due to its high toxicity to arable crops (Emberson and Büker 2008). Ozone concentrations negatively impact natural vegetation (Chuwah et al. 2015) and reduce the ability of ecosystems to mitigate global warming (Sitch et al. 2007) by reducing the CO_2 uptake by plants (Chuwah et al. 2015).Ozone in the troposphere is estimated to rise by 15%

between 2005 and 2050 (Pozzer et al. 2012). Concentration of near-surface ozone is expected to increase significantly through the next century (Sitch et al. 2007).

The Persian Gulf is projected to become an air quality hotspot with rising ozone levels while northern India would experience continuous increase in ozone levels up to 2050 (Pozzer et al. 2012). In East and South Asia, emissions from anthropogenic sources are predicted to be significantly high during 2025–2050 (Pozzer et al. 2012). Greenhouse gas inventory in South Asia, and the sub-Himalaya tract, in particular, has been inadequate (Bhuyan et al. 2014). Therefore, this study assesses the spatial variability of ozone in the Indian Sub-continent during 2002–2016 using AIRX3STM monthly satellite data.

11.2 Database and Methodology

The Atmospheric Infrared Sounder (AIRS) is an instrument aboard the Aqua satellite of National Aeronautics and Space Administration's (NASA). It is an advanced and sophisticated atmospheric sounding system that delivers precise information on the atmosphere's vertical temperature and moisture profile along with specifics of trace greenhouse gases including ozone (https://airs.jpl.nasa.gov/). The AIRX3STM data provides the total integrated column ozone burden in Dobson units (Tian 2017). These gridded datasets have been widely used in recent studies (Rajab et al. 2013; Cui et al. 2013; Gupta et al. 2013; Rafiq et al. 2016; Mahmood et al. 2016).

In the present study, AIRX3STM Level-3 ascending retrieved total integrated column ozone (TotO3-A) Monthly Standard Physical Retrieval (AIRS + AMSU) $1 \times 1^\circ$ spatial resolution V006 datasets were used. Monthly O_3 datasets were downloaded from the NASA website (https://disc.sci.gsfc.nasa.gov/) using the Mirador subsetting tool. A total of 169 NETCDF files were downloaded for the South Asia region. These were imported into a geographic information system as raster files using the multidimensional tools option to facilitate analysis.

The monthly images were combined and averaged to derive month-wise and four season-wise aggregates during 2002–16 in order to understand monthly and seasonal variations in the spatial patterns of O_3.

To assess the overall changes in O_3 during the period 2002–16, a gain coefficient (GC) image was generated following previous studies (Runnstrom 2000).

The regression equation of Y on X is

$$y = a + bx \quad \text{and} \quad b = Sxy/Sxx$$

where $Sxy = \sum (x - \overline{x}) * (y - \overline{y})$ and $Sxx = \sum (x - \overline{x})^2$

To estimate the GC, an average image was derived by aggregating the monthly images over the 2002–16 period to generate the $\overline{y}$-component. Each y ith observation was subtracted from the latter.

The x ith and $\overline{x}$-components were calculated by the use of a spreadsheet program, whereas the GC was derived using the Spatial Analyst and Raster Calculator tools in ArcGIS (www.ersi.com). The GC image was re-classified into five categories, namely, negative, slightly negative, no change, positive, and highly positive. These were further disaggregated for each country.

11.3 The Study Area

The Indian Sub-continent spans an area of 4.4 million km^2 and is home to a population of 1.7 billion. This densely populated land is heavily dependent on agriculture and industries are equally important and growing. With moderate-to-low levels of development prevailing across most of the sub-continent, pollution monitoring is confined to important metropolitan centers. Elsewhere data on pollution levels and at risk populations are scarce. India lacks air quality standards to protect agriculture from ground level ozone (Ghude et al. 2014). The same situation most likely prevails in the rest of the sub-continent, in spite of the entire region being overwhelmingly dependent on agriculture. Developing countries in Asia have less emission controls (Mahmood et al. 2016) in spite of possessing large populations that are vulnerable to air pollution as a health hazard.

Fuel utilization systems, particularly for domestic purposes, are not modern and barring the urban centers are reliant on coal, fuelwood, and biomass. Population and resource requirements, though low per capita, are large considering the sheer size and consistently growing numbers. The area is a prime location for unabated biomass burning and burgeoning fuelwood consumption rates.

The five metropolitan cities in South Asia (Karachi, Mumbai, Delhi, Kolkata, and Dhaka) possess some 15,000 slums with a total population size exceeding 20 million (Weightman 2011). In New Delhi, India's capital city, the innumerable industries and power plants coupled with high per capita vehicular ownership cumulatively add to air pollution (Gurjaret al. 2016). These are aggravated by winter biomass burning in neighboring north Indian states as respiratory illness spiral. The pollution levels peak dangerously high during winter, but solutions are short term and impermanent and recur every year. Dhaka is among the most polluted cities in the world. Governmental efforts to reduce lead pollution in the face of population and urban growth, natural calamities, and political or other instabilities or inefficiencies tend to slow or halt implementation (Weightman 2011). Pollution in megacities and other areas is responsible for over 628,000 premature deaths every year in India, making it the fifth leading cause of premature mortality (Gurjaret al. 2016). Megacity plumes contain large amounts of different pollutants including greenhouse gases, ozone precursors, and aerosols (Gurjaret al. 2016).

Temperature inversions are common in Kathmandu since it is located in a trough like hollow and is hemmed in by mountains that surround it (Weightman 2011). As a result, the air is polluted by dust, smog, vehicular emissions from cars and buses, and dirt spewed out by brick and carpet factories (Weightman 2011). In fact during winter substantial quantities of dust are blown in by monsoonal winds out of India (Weightman 2011). The Indo-Gangetic Plain (IGP), which is among the most important fertile agricultural belt, experiences high surface O_3 concentrations (Engardt 2008).

Disconcertingly, ozone concentrations are projected to accentuate in the coming years (Avnery et al. 2011), an occurrence that would detrimentally affect the agricultural sector's vulnerability (Ghude et al. 2014).

Surface O_3 in India is expected to increase in future considering the recent upward trend in precursor emissions (Ghude et al. 2014). India suffers from some of the worst air pollution in the world, owing to its rapid economic development, increasing population, growth in energy demand, and limited air-pollution regulation (Karambelas et al. 2018).

11.4 Results

Monthly averaged imageries (2002–16) show high O_3 concentration centered in northeast Pakistan that remains consistent throughout the year barring the summer months of July–August. From January to June, it remains centered on northeast Pakistan and deviates to northwest India during March–April (Fig. 11.1). During January–March the rest of the sub-continent registers lower levels of O_3 ranging from a minimum of 245–267 DU to a high of 280–280 DU. The Himalaya, Bangladesh, and Sri Lanka including peninsular India and south and southwest Pakistan register moderate-to-low O_3 levels. In April–May, the O_3 hotspot shifts eastwards from NE Pakistan into NW India and to the western portion of the Indo-Gangetic plain (IGP). However, this situation does not persist and by June O_3 levels decline. In June, the O_3 levels are low in general with only a few areas with O_3 concentrations above 300 DU (Figs. 11.1 and 11.2).

During November and December, the highest levels of O_3 dip below 295 DU and the lowest O_3 levels reach as low as 251 DU. A similar weakened O_3 scenario persists during November–December when a small area in NW Pakistan starts to show an O_3 concentration. The level of O_3 however remains well below the 300 DU mark, hovering around 283–289 DU. High O_3 levels over Tibet in winter, from October till March, accrued as intruding from tropospheric air passed over snow clad mountain surfaces (Engardt 2008). A net decrease in tropospheric ozone concentration over NE during the period 1979 to 2009 has been observed (Kalita and Bhuyan 2011). The highest concentrations of O_3 during 2000 accrued in north India, north Pakistan, and some areas of Bangladesh (Engardt 2008). High concentrations in the Ganga Valley resulted from high and substantive precursor emissions in the area (Engardt 2008).

To assess the seasonal variation of O_3, the raster images were combined season-wise for the summer months of June, July, and August (JJA); autumn/late summer months of September, October, and November (SON); winter months of December, January, and February (DJF); and the spring months (March, April, and May). The

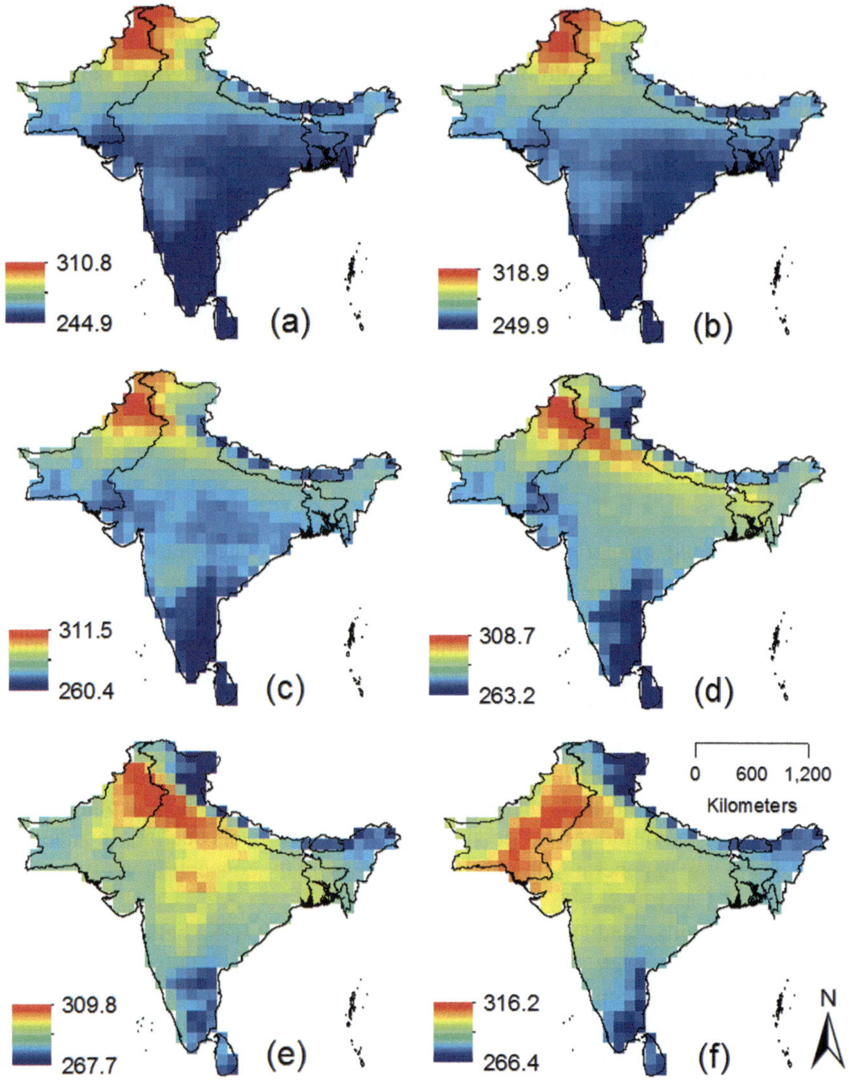

Fig. 11.1 Monthly averaged O_3 imageries (2002–16) during **a** January, **b** February, **c** March, **d** April, **e** May, and **f** June indicate the spatial shift of the ozone hotspot

seasonal aggregates were assessed for two trienniums during 2002–03 to 2005–06 and 2013–14 to 2015–16 to discern changes therein. The season-wise were then averaged in a GIS using the raster calculator function. During JJA a similar situation exists during 2003–05 and 2014–16 (Fig. 11.3, inset a and b) wherein the highest levels of O_3 remain confined to Pakistan and NW India.

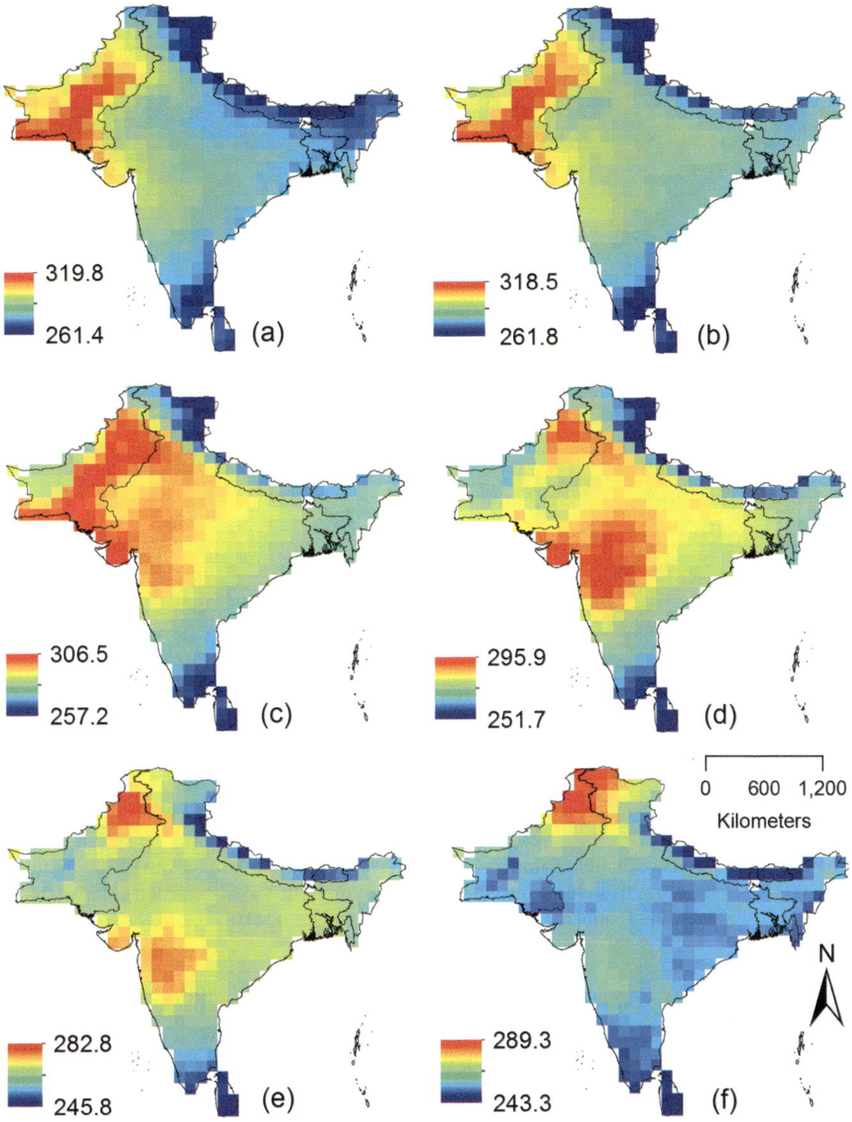

Fig. 11.2 During July–August (inset **a**) and (**b**) a southward shift from NW Pakistan to SE Pakistan took place. Also the areal extent with pixels above 300 DU increased during August as much of Pakistan shows high O_3 levels. During September–October the high O_3 belt shifts into western and central India (inset **c**) and (**d**). A slight shift and decline is evident during November (inset **e**) and December (inset **f**). During December the high concentration area is at its minimum and is confined to the NW corner of the sub-continent

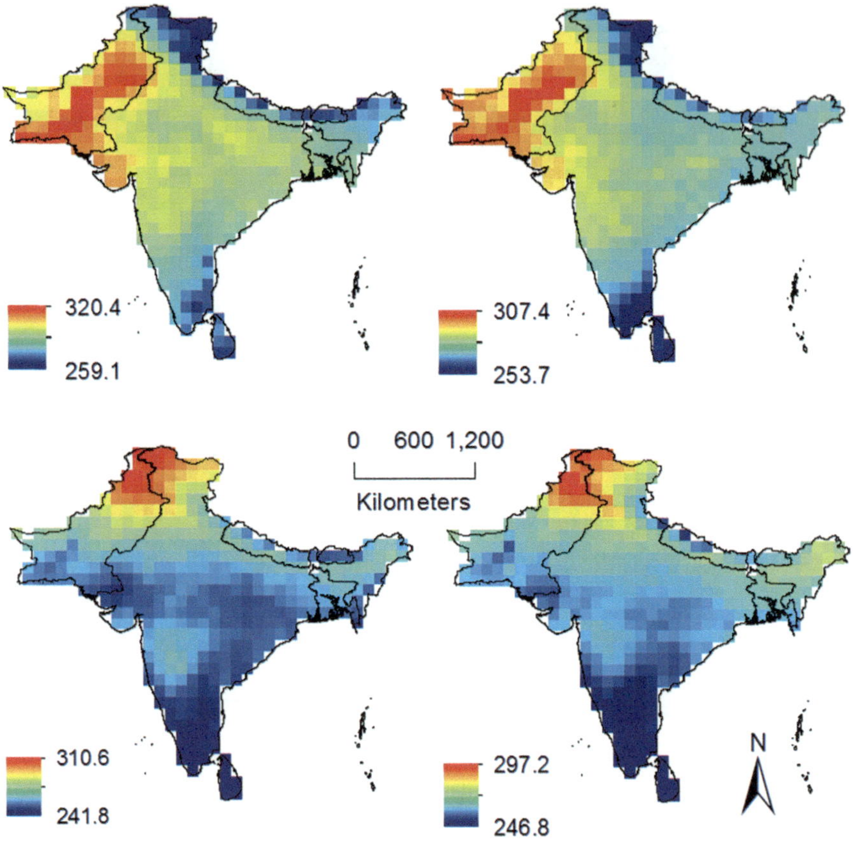

Fig. 11.3 Season-wise trienniums during 2002–03 to 2005–06; and 2013–14 to 2015–16 during JJA 2003–05 (top left); 2014–16 (top right) and DJF during 2003–05 (lower left); and 2014–16 (lower right) indicate the spatial seasonal shift and a decline during DJF

A 4.2% decline in the range of values is, however, discernible from 320 DU in 2003–05 to 307 in 2014–16. The lower limits of O_3 also show a decline over the same period.

During winter a similar decrease in O_3 levels is seen when DJF values show a 4.3% reduction from 310 DU during 2002–04 to 297 DU during 2013–2015. During this period the O_3 hotspot seems to shrink in size from a larger spatial extent to one of a smaller area centered on NE Pakistan and NNW India (Fig. 11.3, lower left and right).

In terms of spatial extent, an increase in area showed for O_3 levels over 300 DU during MAM (Fig. 11.4). From a small area in NE Pakistan, it expanded to NW India during 2003–05 and 2014–16.

Similar trends occurred during SON in terms of spatial extent wherein a larger area showed higher O_3 levels. During SON no such decline in O_3 intensity occurred.

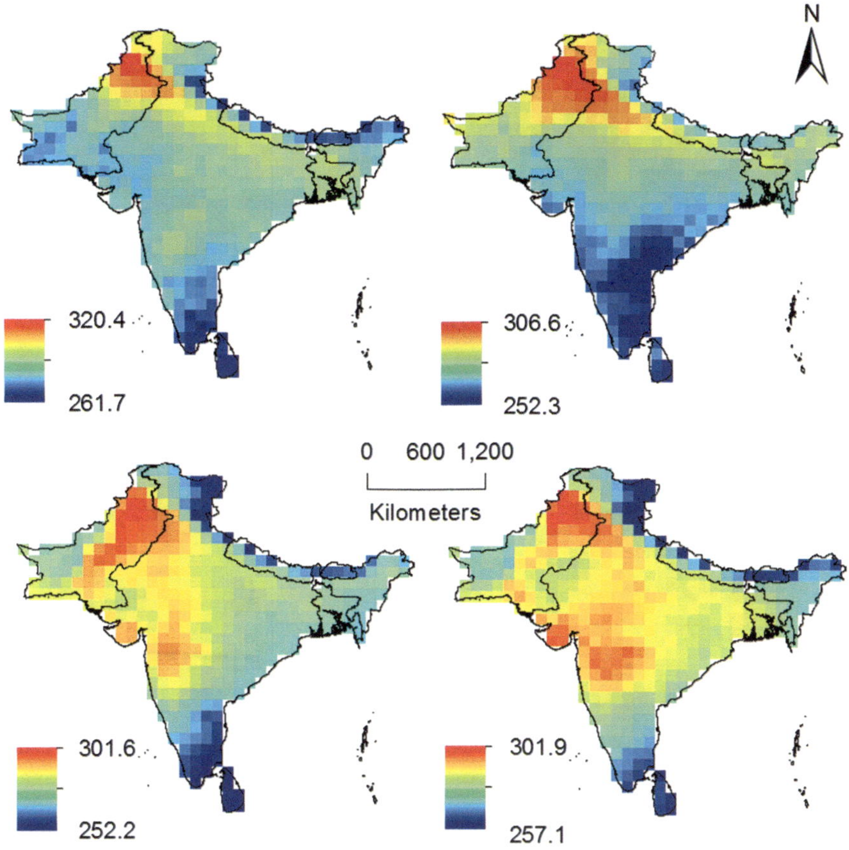

Fig. 11.4 Season-wise trienniums during 2002–03 to 2005–06 and 2013–14 to 2015–16. MAM is shown during 2002–03 to 2005–06 (above, left) and 2013–14 to 2015–16 (above, right); SON 2002–03 to 2005–06 (lower, left) and during 2013–14 to 2015–16 (lower, right) reveal seasonal variations

While at the upper end of the values no changes occurred during SON, at the lower end of the O_3 values, a marginal increase from 252 to 257 DU took place (Fig. 11.4).

To assess the changes in O_3 during 2001–16 in the Indian Sub-continent a gain coefficient image was generated (Fig. 11.5).

The highly positive areas lie in a narrow stretch of land along the Pakistan–Afghanistan border on the one hand and on the other on an elongated strip of land skirting the Himalaya stretching from the northeast of New Delhi and passing through Nepal, Bhutan, and the tip of NE India bordering eastern Bhutan. These two stretches of positive gain account for 18.5% of the total geographical area of the Indian Sub-continent. The NW Pakistan O_3 hotspot which had consistently shown high O_3 levels throughout the 2002–16 period and across most seasons did not show a positive

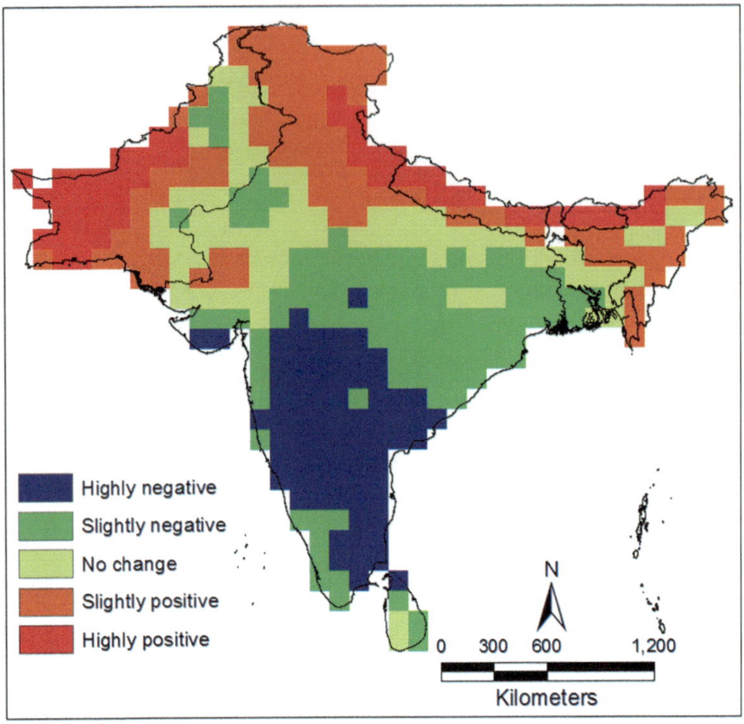

Fig. 11.5 The gain coefficient image showing the changes in O_3 during 2002–16 in the Indian Sub-continent

gain coefficient. Biomass burning was correlated with tropospheric ozone concentrations over northeast India and the IGP (Kalita and Bhuyan 2011). In the former, lightning incidence during spring was also a determining factor behind high ozone levels (Kalita and Bhuyan 2011). Biomass burning includes the effects of shifting cultivation practices in northeast India wherein forest areas are cleared for cultivation in hill areas using fire and a rotation of fields is the norm. This is in contrast to north and northwest India where biomass residues left in the harvested paddy fields are burnt during early winter. Additionally, dung, which, when not used as a fertilizer, is used as cooking fuel; such dung cooking fires being major sources of pollution (Weightman 2011). In rural areas of the sub-continent, primary particulate emissions from residential combustion sectors are common in rural and low-income urban regions, where people rely more on traditional biomass to meet their cooking and heating needs (Karambelas, et al. 2018). In higher elevation areas of the Indian Sub-continent, such as the Himalaya and its foothill zones during the cold season, biomass and fuelwood are burnt for space heating purposes exacerbating pollution levels.

11.5 Conclusions

Tropospheric O_3 emissions over the Indian Sub-continent were examined using satellite datasets during 2002–16. AIRX3STM Level-3 ascending retrieved total integrated column ozone (TotO3-A) Monthly Standard Physical Retrieval (AIRS + AMSU) $1 \times 1°$ spatial resolution V006 datasets were used in the present analysis.

O_3 emissions indicated a substantial spatial heterogeneity over the Indian Sub-continent. Increases in tropospheric ozone emissions are not unique to the Indian Sub-continent; in fact, these emissions are projected to increase in several countries across the globe. Projections suggest that climate change will increase concentrations of tropospheric ozone, at least in high-income countries when precursor emissions are held constant, which would increase morbidity and mortality (Ebi and McGregor 2008). However, implementation of stringent air-pollution policies could lead to a reduction in episodic peak ozone concentrations in Europe and North America (Chuwah et al. 2015). In fact, regulatory interventions to reduce air pollution in the United States and Europe have seen substantial improvements in recent decades (Jerrett 2015; Anderson et al. 2016). However, such improvements have not accrued in the tropics, which continue to exert an adverse climatic impact of O_3, namely, its radiative influence (Anderson et al. 2016).

Surface ozone concentrations in regions with rapidly growing economies in South and East Asia are projected to increase at least in the near term (Chuwah et al. 2015). O_3 concentration is on the rise in the South Asian region as a result of growing anthropogenic precursor emissions (Engardt 2008). Such a scenario is compounded by poor pollution reduction implementation mechanisms in the Indian Sub-continent. The overlap of high pollution and large populations takes a huge toll on public health in India and other Asian countries (Jerrett 2015).The situation is likely to worsen emissions of several key pollutants in these countries which is expected to increase in the future (Jerrett 2015). Significant increase in the risk of deaths from respiratory causes due to raised ozone concentrations (Jerrett et al. 2009) holds serious implications for the Indian Sub-continent; since India, Pakistan, and Bangladesh show among the highest population-weighted ozone level increases in the world during 1990–2015 (Cohen et al. 2017). Certain studies have projected that mortality estimates from air pollution could double by 2050 based on spiraling population and pollution rates (Jerrett 2015). Thus, regulatory authorities in countries that comprise the Indian Sub-continent as well as governmental agencies need to address these issues on a priority basis.

References

Anderson DC, Nicely JM, Salawitch RJ, Canty TP, Dickerson RR, Hanisco TF, Wolfe GM, Apel EC, Atlas E, Bannan T, Bauguitte S, Blake NJ, Bresch JF, Campos TL, Carpenter LJ, Cohen MD, Evans M, Fernandez RP, Kahn BH, Kinnison DE, Hall SR, Harris NRP, Hornbrook RS, Lamarque J-F, Le Breton M, Lee JD, Percival C, Pfister L, Pierce RB, Riemer DD, Saiz-Lopez A,

Stunder BJB, Thompson AM, Ullmann K, Vaughan A, Weinheimer AJ (2016) A pervasive role for biomass burning in tropical high ozone/low water structures. Nat Commun 7:10267. https://doi.org/10.1038/ncomms10267

Avnery S, Mauzerall DL, Liu J, Horowitz LW (2011) Global crop yield reductions due to surface ozone exposure: 1. Year 2000 crop production losses and economic damage. Atmos Environ 45:2284–2296

Bhuyan PK, Bharali C, Pathak B, Kalita G (2014) The role of precursor gases and meteorology on temporal evolution of O_3 at a tropical location in northeast India. Environ Sci Pollut Res 21(10):6696–6713

Bourgeois I, Peischl J, Andrew Neuman J, ... Wofsy SC, Ryerson TB (2021) Large contribution of biomass burning emissions to ozone throughout the global remote troposphere. Proc Nat Acad Sci USA 118(52):e2109628118

Chuwah C, van Noije T, van Vuuren DP, Stehfest E (2015) Global impacts of surface ozone changes on crop yields and land use. Atmos Environ 106:11–23

Cohen AJ, Brauer M, Burnett R, Anderson HR, Frostad J, Estep K, Balakrishnan K, Brunekreef B, Dandona L, Dandona R, Feigin V, Freedman G, Hubbell B, Jobling A, Kan H, Knibbs L, Liu Y, Martin R, Morawska L, Pope CA, Shin H, Straif K, Shaddick G, Thomas M, van Dingenen R, van Donkelaar A, Vos T, Murray CJL, Forouzanfar MH (2017) Estimates and 25-year trends of the global burden of disease attributable to ambient air pollution: an analysis of data from the Global Burden of Diseases Study 2015. Lancet 389:1907–1918

Cui Y, Du J, Zhang D, Sun Y (2013) Anomalies of total column CO and O_3 associated with great earthquakes in recent years. Nat Hazard 13:2513–2519

Ebi KL, McGregor G (2008) Climate change, tropospheric ozone and particulate matter, and health impacts. Environ Health Prespect 116, 11:1449–1455

Emberson L, Büker P (2008) Ozone: a threat to food security in South Asia. Stockholm Environmental Institute, Policy Brief, Stockholm.

Engardt M (2008) Modelling of near-surface ozone over South Asia. J Atmos Chem 59:61–80

Ghude SD, Jena C, Chate DM, Beig G, Pfister GG, Kumar R, Ramanathan V (2014) Reduction in India's crop yield due to ozone. Geophys Res Lett 41:5685–5691

Gupta A, Kumar V, Panwar V, Bhatnagar R, Dhaka SK (2013) Long term variability in temperature in the upper troposphere and lower stratosphere over Indian and Indonesian region using Atmospheric Infrared Sounder (AIRS) observations. Indian J Radio Space Phys 42:298–308

Gurjar BR, Ravindra K, Nagpure AS (2016) Air pollution trends over Indian megacities and their local-to-global implications. Atmos Environ 142:475–495

Hanslmeier A (2008) The sun and space weather. Springer, Dordrecht

Jerrett M, Burnett RT, Arden Pope III C, Ito K, Thurston G, Krewski D, Shi Y, Calle E, Thun M (2009) Long-term ozone exposure and mortality. New Engl J Med 360:1085–1095

Jerrett M (2015) The death toll from air-pollution sources. Nature 525:330–331

Kalita G, Bhuyan PK (2011) Spatial heterogeneity in tropospheric column ozone over the Indian subcontinent: long-term climatology and possible association with natural and anthropogenic activities. Adv Meteorol 924516, 12 pp. https://doi.org/10.1155/2011/924516

Karambelas A, Holloway T, Kiesewetter G, Heyes C (2018) Constraining the uncertainty in emissions over India with a regional air quality model evaluation. Atmos Environ 174:194–203

Kondo Y, Morino Y, Takegawa N, ... Streets DG, Liley B (2004) Impacts of biomass burning in Southeast Asia on ozone and reactive nitrogen over the western Pacific in spring. J Geophys Res D: Atmos 109(15)

Lelieveld J, Evans JS, Fnais M, Giannadaki D, Pozzer A (2015) The contribution of outdoor air pollution sources to premature mortality on a global scale. Nature 525(7569):367–371

Mahmood I, Iqbal MF, Shahzad MI, Waqas A, Luqman A (2016) Spatiotemporal monitoring of CO_2 and CH_4 over Pakistan Using Atmospheric Infrared Sounder (AIRS). Int Lett Nat Sci 58:35–41

Platt U, Poschl U, …Rudich Y, Sellegri K, Steinbrecher R, Simpson D, ten Brink H, Theloke J, van der Werf GR, Vautard R, Vestreng V, Vlachokostas C, von Glasow R (2009) Atmospheric composition change—global and regional air quality. Atmos Environ 43:5268–5350

Pozzer A, Zimmermann P, Doering UM, van Aardenne J, Tost H, Den tener F, Janssens-Maenhout G, Lelieveld J (2012) Effects of business-as-usual anthropogenic emissions on air quality. Atmos Chem Phys 12:6915–6937
Rafiq L, Tajbar S, Manzoor S (2016) Long term temporal trends and spatial distribution of total ozone over Pakistan. Egypt J Remote Sens Space Sci 20:295–301
Rajab JM, Lim HS, Matjafri MZ (2013) Monthly distribution of diurnal total column ozone based on the 2011 satellite data in Peninsular Malaysia. Egypt J Remote Sens Space Sci 16(1):103–109
Roaf S, Crichton D, Nicol F (2009) Adapting buildings and cities for climate change a 21st century survival guide, 2nd edn. Elsevier, Burlington
Runnstrom M (2000) Is northern China winning the battle against desertification? Satellite remote sensing as a tool to study biomass trends in the Ordos Plateau in semiarid China. Ambio 29:468–476
Sauvage B, Martin RV, van Donkelaar A, Ziemke JR (2007) Quantification of the factors controlling tropical tropospheric ozone and the South Atlantic maximum. J Geophys Res 112:D11309
Sharma A, Ojha N, Pozzer A, Mar KA, Beig G, Lelieveld J, Gunthe SS (2017) WRF-Chem simulated surface ozone over south Asia during the pre-monsoon: effects of emission inventories and chemical mechanisms. Atmos Chem Phys 17, 23:14393–14413
Sitch S, Cox PM, Collins WJ, Huntingford C (2007) Indirect radiative forcing of climate change through ozone effects on the land-carbon sink. Nature 448:791–795
Tian B (2017) AIRS/AMSU/HSB version 6 level 3 product user guide. Jet Propulsion Laboratory California Institute of Technology, Pasadena
Wang YH, Liu SC, Wine PH, Davis DD, Sandholm ST, Atlas EL, Avery MA, Blake DR, Blake NJ, Brune WH, Heikes BG, Sachse GW, Shetter RE, Singh HB, Talbot RW, Tan D (2001) Factors controlling tropospheric O_3, OH, NOx and SO_2 over the tropical Pacific during PEM-Tropics B. J Geophys Res 106:32733–32747
Weightman BA (2011) Dragons and tigers: a geography of South, East, and Southeast Asia, 3rd edn. Wiley

Chapter 12
Geo-Environmental Problems in the Fringe Areas of the Rudrasagar Lake, Tripura, India

Istak Ahmed, Nibedita Das (Pan), Jatan Debnath, and Haradhan Banik

Abstract The Rudrasagar Lake is a famous tourist spot in northeast India owing to the presence of the Neermahal Palace within it. Unfortunately, growing human intervention within its catchment area has led to its gradual degradation and areal shrinkage. Thus the current analysis seeks to assess the various geo-environmental problems that accrue in the area surrounding the Rudrasagar Lake and their impacts. Four cross sections were taken across the connecting channels. Land use land cover (LULC) change within a 1 km buffer zone around this lake over a 30-year period (1988–2018) was analyzed. The USLE model was used to identify the nature of soil loss from the lake catchment. A household survey was conducted to identify the issues stemming from the degradation of the lake. The study reveals a significant increase in settled and agricultural land within the immediate surroundings of this lake during the period 1988–2018. Moreover, the area under moderate, high, very high, and severe soil loss within the Rudrasagar Lake catchment has increased manifold within the last few decades. It has accelerated the problem of siltation in this lake. Nearly 89% of the surveyed respondents felt that fish production from this lake had gradually decreased causing a section of the local fishermen to shift their occupation partially or completely.

Keywords Lake · Soil loss · Sediment load · Eutrophication · Rudrasagar Lake

12.1 Introduction

Lakes are a unique feature of the earth's surface and extremely valuable ecosystems that provide a wide range of services to mankind (Madhumathi et al. 2018). Some examples of such life-supporting services are the regulation of local/global hydrological and climatic regimes, provision of water for drinking, electricity generation, fisheries, irrigation, recreation, etc. (Durmaz et al. 2016). However, increased human

---

I. Ahmed (✉) · N. Das (Pan) · J. Debnath · H. Banik
Department of Geography and Disaster Management, Tripura University, Suryamaninagar, Tripura 799022, India
e-mail: Istak223311@gmail.com

A. Saikia and P. Thapa (eds.), *Environmental Change in South Asia*,
https://doi.org/10.1007/978-3-030-47660-1_12

intervention in and around lakes is leading to regular modification of their environments. Urbanization, pollution from different sources including sewage, agricultural runoff rich with nutrients, industrial toxic liquid waste, and reclamation are major factors that lead to the gradual degradation of lakes (Saraf and Nair 2015). Land use of the catchment area of lakes is subject to regular alteration through clearance of natural vegetation in order to cater to various human needs. Moreover, encroachment on lake shores through sewage disposal and various developmental activities is also emerging as a matter of concern. All these geo-environmental problems accelerate the rate of sediment supply which in turn led to gradual siltation and shrinkage of the lake.

Tripura, a small state in northeast India is blessed with a large number of lakes, the Rudrasagar being one such lake. It was identified as a Ramsar site on November 8, 2002 attesting to its international importance. This lake provides a potential natural habitat to many bird species and is also important for many endemic fish species (Choudhury 2008; Barman et al. 2013). Unfortunately, this lake is prone to severe anthropogenic disturbances in the form of overfishing, pollution, agricultural encroachment, etc. (Maurya et al. 2020). These activities have cumulatively and consistently degraded its natural environment. Thus, the present analysis aims to assess the problems prevailing in and around the Rudrasagar Lake and their detrimental impacts so that proper strategies can be formulated to better manage this Ramsar lake and its fringe area.

12.2 The Study Area

The Rudrasagar Lake (23°29′N and 90°01′E) is located in the Melaghar Block of Sepahijala District and is situated around 52 km away from the state capital of Tripura (Fig. 12.1). The Neermahal Palace built during 1930–1938 is located in the middle of this lake which makes it an important tourist destination. The then King of Tripura, ("Maharaja" was the term given to Kings in India) Maharaja Bir Bikram Kishore Manikya Debbarma of the erstwhile Kingdom of Tripura, built the palace. The lake is basically a wetland located within the flood plain of the Gumti River basin. The Rudrasagar Lake receives an inflow of water from three perennial streams, namely, Noa Cherra, Kemtali Cherra, and Durlavnarayan Cherra and outflow is discharged into the Gumti River through a channel, Kachigang (Abir 2014; Pal et al. 2015, 2016). The bed of this lake is formed of silt deposition which is mainly received from the connecting streams that provide it with a natural supply of fresh water. The surroundings of the lake are mainly covered by two physiographic units, namely flood plains and hillocks or terraces and or uplands (locally called a "tila"), respectively. This lake attracts the attention of geo-scientists and environmentalists considering its physical and ecological significance.

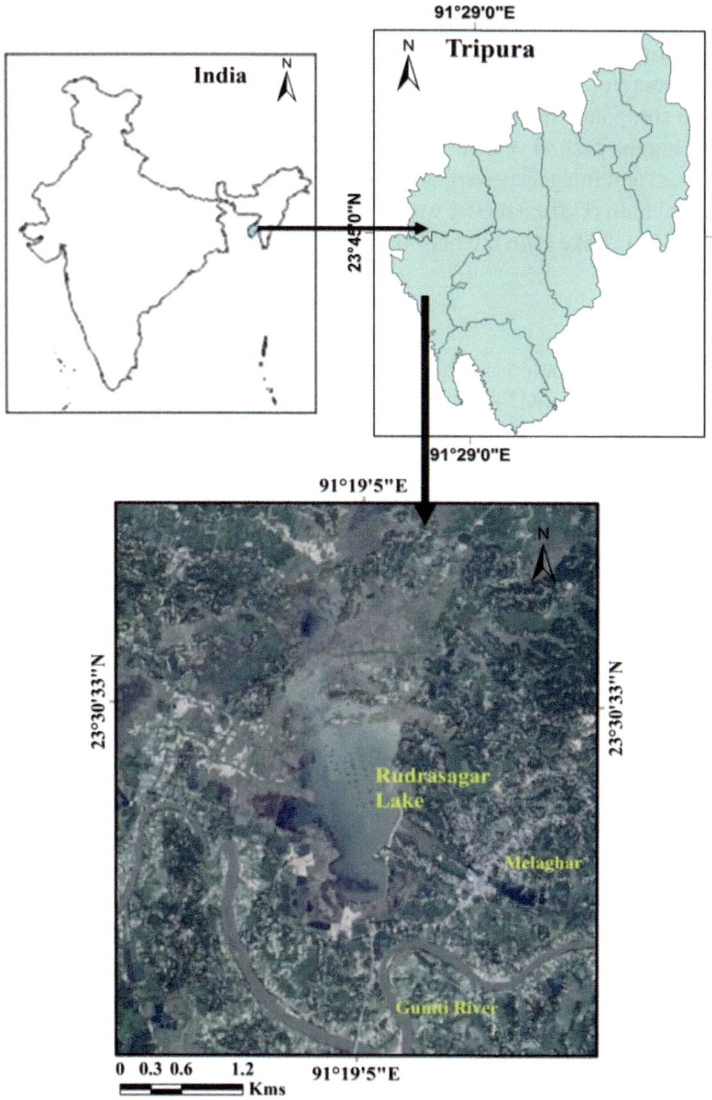

Fig. 12.1 Location of the study area

12.3 Material and Methods

The present study was carried out on the basis of field visits and modern geospatial technology. During fieldwork, two cross sections were taken across the stream Noa Cherra and another two across Kachi Gang. Besides, water samples were also collected from these four sites to measure the suspended sediment concentration. The motive was to assess the variation in amount of sediment supply to the Rudrasagar Lake and that released from it. A one km buffer zone, surrounding the Rudrasagar Lake, was generated using a geographic information system, Arc GIS 10.1 (www.esri.com) to study land use land cover status of this zone and to detect changes that accrued during the past 30 years (1988–2018). Landsat satellite imagery of 1988 and 2018 was used and imageries were accessed from the USGS website. The Universal Soil Loss Equation (USLE) model was used to measure soil loss from the catchment of the Rudrasagar Lake with the help of following formula:

$$A = R.K.LS.C.P$$

where A denotes computed average annual soil loss (ton^{-1} ha^{-1} year), R is the rainfall erosivity factor (MJ mm ha^{-1} h^{-1} $year^{-1}$), K is the soil erodibility factor (t MJ^{-1} yr^{-1} ha^{-1} h mm^{-1}), LS (dimensionless) is the topographical factor (slope length and slope steepness factor), C (dimensionless) is crop management factor and P is conservation practice factor.

The water area of the Rudrasagar Lake was obtained from satellite images of different years (1988, 2003, and 2018) and superimposed to get an insight about the spatial extent of this lake. Images during January were used as there was less possibility of an extra influx of water in the form of rainfall during this month and also cloud cover during winter is not problematic in tropical and near tropical areas. Data regarding the number of tourists that visited the Neermahal Palace during the period 2009–2017 was collected from the Department of Tourism, Government of Tripura. Satellite imagery of October 25, 2017 was used to identify the inundated zone in the surroundings of the water area of the Rudrasagar Lake. The extracted polygon of this zone was converted into a KML file and imported into Google Earth to prepare a land use map of the area. Moreover, different households of selected villages, located along this lake were surveyed to gather primary data on various environmental issues related to this lake. The targeted age group of respondents was aged $50–75^{+}$ years so that we could derive an assessment from respondents who had first-hand experience of the changes that had taken place in and around the Rudrasagar Lake over the past few decades. The data and information collected from the field were then arranged in tabular form and represented graphically.

12.4 Results and Discussion

12.4.1 Geo-Environmental Problems

12.4.1.1 Population Pressure

The population around the Rudrasagar Lake started to increase when refugee colonies were established in villages surrounding the lake. In the year 1951, there were

Table 12.1 Population status of five villages located in the vicinity of the Rudrasagar Lake

Name of the village	Population		Population increase (%)
	(1991)	(2011)	
Paschim Nalchar	4,384	7,887	44.41
Rudijala	4,202	5,856	28.24
Durlavnarayanpur	4,063	5,819	30.17
Choumohani	6,188	8,317	25.60
Khas Choumohani	1,854	2,119	12.51

Source Census of India, 1991 and 2011

only 600 families who migrated from Cumilla and Brahmanbaria of East Pakistan (presently Bangladesh) into Tripura. After a few years, the population started to gradually increase because of an increase in agricultural activities and practice of fishing and further ingress of population.

From 1991 to 2011, the maximum increase was observed in Paschim Nalchar village where the total population increased by 44.41% (Table 12.1). In the four other villages, it increased by 12.51–30.17%. Such population increases generated significant pressure on the surrounding natural resources and resulted in the alteration of the lake's natural environment.

12.4.1.2 Changing Trends in Tourist Footfalls

Given the presence of the spectacular and majestic Neermahal Palace located in the middle of the Rudrasagar Lake, this is a popular tourist destination. A large number of tourists from India and abroad visit it every year.

In the financial year 2014–15, number of tourists' visit to Neermahal was slightly lesser than in the other years (Fig. 12.2) due to the ongoing renovation work of the palace. Whereas in the next two consecutive financial years (2015–16 and 2016–17) it increased considerably, especially during 2016–17 when the number of tourists increased to 108,839. The tourists tend to dispose plastic bags, bottles, plates, etc. at picnic spots on the banks of the Rudrasagar Lake (Fig. 12.3). Such wastes not only contaminate the water of the lake but also result in its physical reduction through the accumulation of these materials. The lack of proper disposal facilities at the lake site is indeed a problematic issue.

12.4.1.3 Changing Pattern of Soil Loss in the Catchment of the Rudrasagar Lake

The catchment of the Rudrasagar Lake consists of two sub-catchments, namely, Noa Cherra and Durlavnarayan-Kemtali Cherra. Over the last few decades, this catchment has experienced several changes that have impacted the total amount of soil loss. USLE and Revised Universal Soil Loss Equation (RUSLE) have been widely used

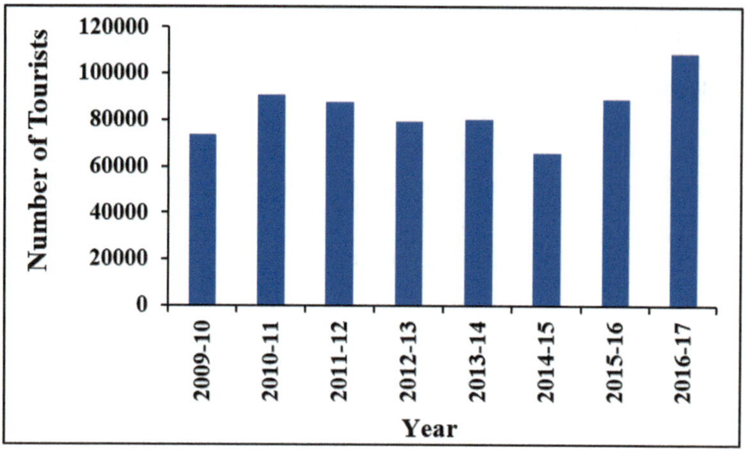

Fig. 12.2 Number of tourists visiting the Neermahal Palace (2009–2017). *Source* Department of Tourism, Government of Tripura

Fig. 12.3 Disposal of waste materials by the tourists in Rudrasagar Lake

to predict soil loss because of their convenience in application and compatibility with GIS (Dabral et al. 2008; Prasannakumar et al. 2012). In the present analysis, the USLE model was applied to assess the changing pattern of soil loss from this catchment during the period 1990–2016 (Figs. 12.4 and 12.5). The areas under the severe soil loss category show less variation during that period in comparison to the other categories. However, the fact remains that a significant area under the low soil loss category got converted from moderate to high soil loss category.

The area under the low soil loss category decreased from 141.2 to 130.06 km^2 from 1990 to 2016 (Table 12.2). During the same time, the area under the moderate soil loss category increased from 28.59 to 34.58 km^2 and the area under the high soil loss category grew four-fold from 1.02 to 4.09 km^2. These results attest to the fact that the rate of erosion in the Rudrasagar Lake catchment is gradually increasing

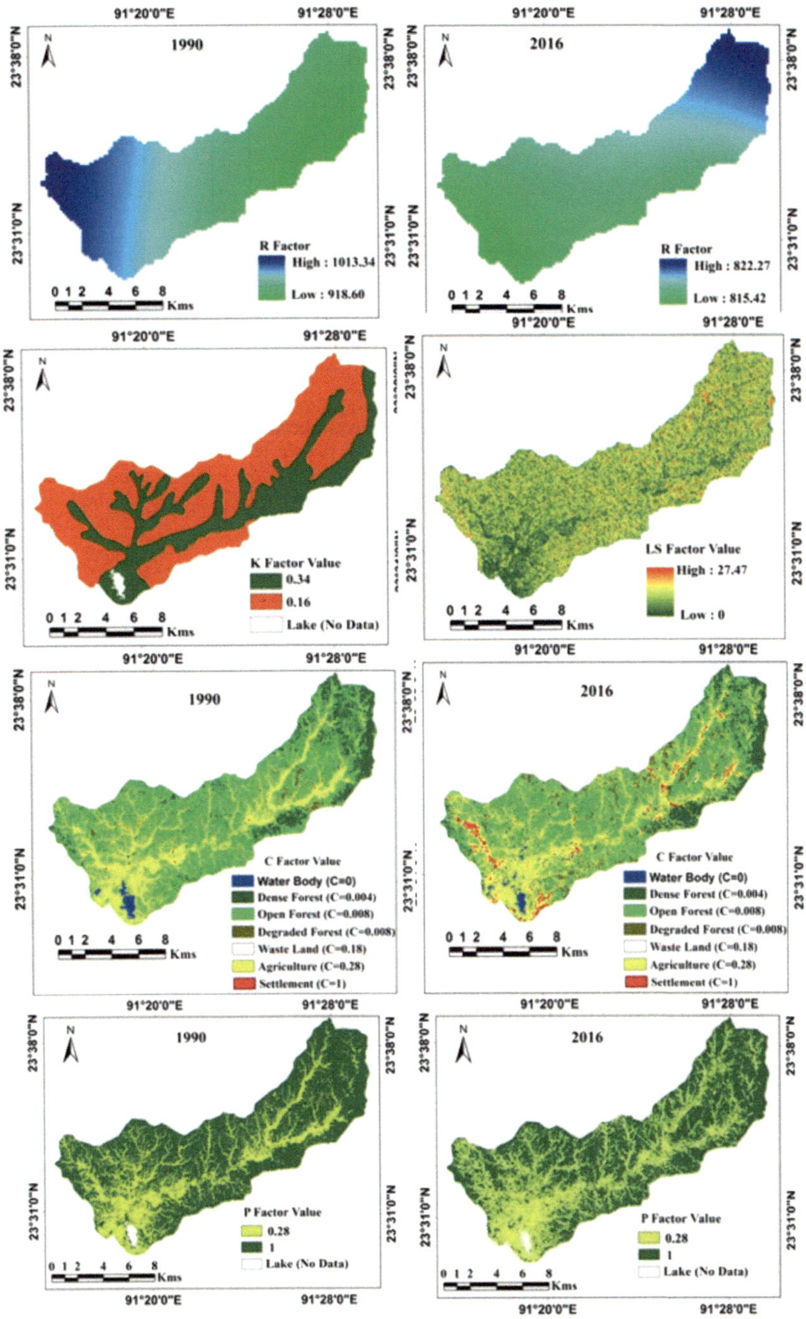

Fig. 12.4 Maps showing different parameters of USLE model of the Rudrasagar Lake catchment for the years 1990 and 2016

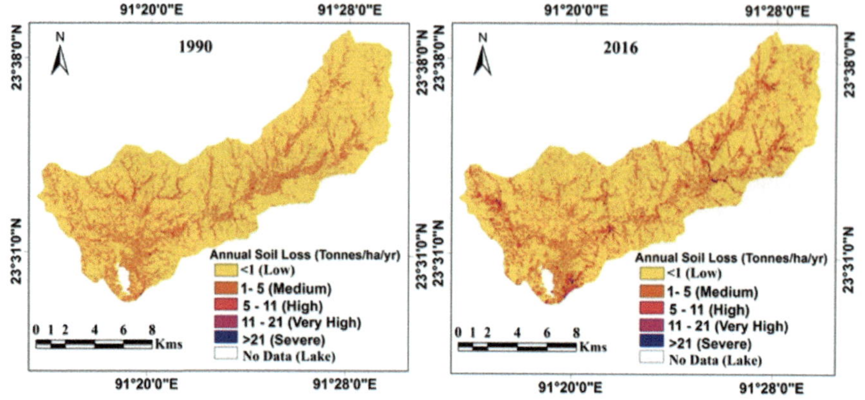

Fig. 12.5 Soil losses from the Rudrasagar Lake catchment between 1990 and 2016

Table 12.2 Areal differentiation in various soil loss categories (1990–2016)

Amount of soil loss (Tons /ha/yr)	Area (Km^2)		Change (in %)[a]
	1990	2016	
Below 1 (Low)	141.2	130.06	−8.9
1–5 (Moderate)	28.59	34.58	+20.95
5–11 (High)	1.02	4.9	+380.39
11–21 (Very High)	0.39	1.5	+284.61
Above 21 (Severe)	0.13	0.45	+246.15

[a]Decrease is denoted by—and increase by +sign

and the extra eroded materials are directly or indirectly adding up to the previous deposits in the lake resulting in a gradual shrinkage to its areal extent over time.

12.4.1.4 Sluice Gate Across the Kachi Gang and Its Effect

Water that discharges from the Rudrasagar Lake through Kachigang is regulated by a sluice gate before its confluence with the Gumti River. It controls the backflow of water from this river, especially during the rainy season. It has been observed from the cross sections that both the suspended sediment concentration and sediment discharge are highly variable in the waters that enter the Rudrasagar Lake through the Noa Cherra and the water that is released from the lake through Kachigang (Fig. 12.6).

Sediments noticeably decreased in water released from the lake (Fig. 12.7). A decrease in the suspended sediment discharge implies that more sediments were getting accumulated within the lake and this is not a healthy signal for the Rudrasagar lake in the long term. This fact is also supported by variation in the visual appearance

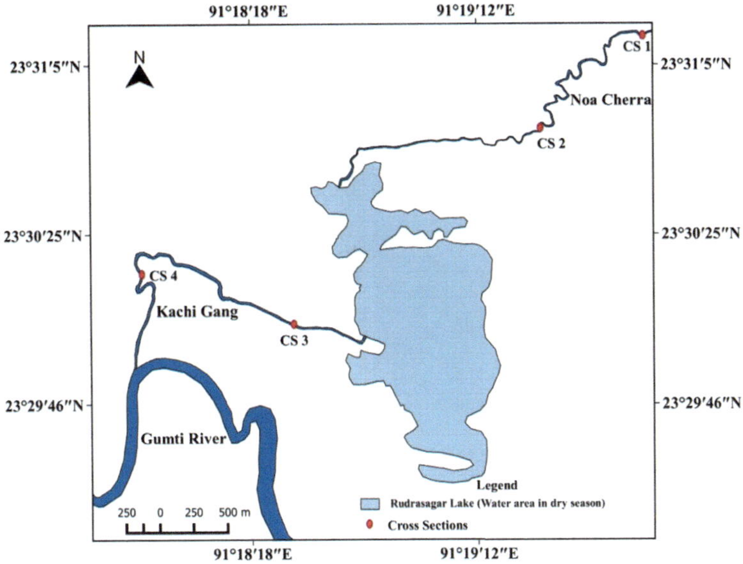

Fig. 12.6 Location of cross sections taken across the Noa Cherra and Kachigang

of the water color in Google Earth imagery (Fig. 12.8). It reveals that a significant amount of sediment brought by the Noa Cherra gets deposited within the Rudrasagar Lake which over the years has resulted in the silting up of its bed as water released from this lake is unable to move freely due to the presence of the sluice gate across Kachi Gang.

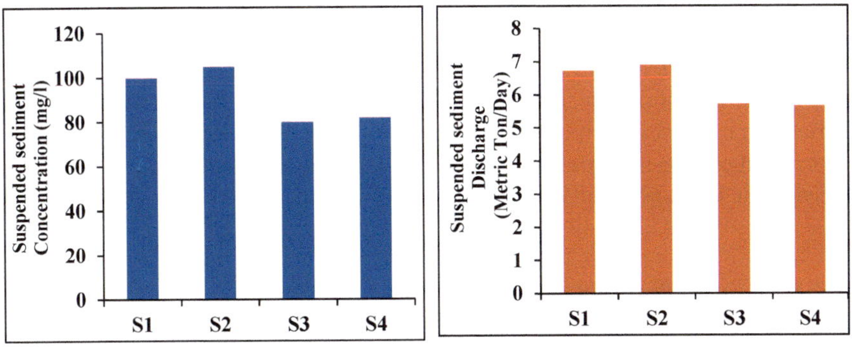

Fig. 12.7 Variation of suspended sediment concentration and discharge in water added to the Rudrasagar Lake (CS1 and CS2) and released from it (CS3 and CS4)

Fig. 12.8 Physical appearance of water added and released from the Rudrasagar Lake (*Source* Google Earth Imagery)

12.4.1.5 Changes in Land Use/Landcover

Growing population pressure in the surroundings of the Rudrasagar Lake has resulted in the alteration of land use/land cover over the area. A 1 km buffer zone was created to assess the change in land use/land cover pattern from the year 1988 to 2018 (Fig. 12.9). The area under natural vegetation within this buffer zone was 478.62 acres in the year 1988, which decreased to 270.26 acres by 2018. At the same time, the area under settlement increased from 68.65 to 178.46 acres between 1988 and 2018, respectively (Fig. 12.10). Such an increase in the settled area was marked mainly in the eastern part under the Melaghar Municipal Council. During this period, significant growth in the agricultural area was also observed. Although the actual lake area reduced during that period, at the same time, the area under the water body increased marginally mainly due to the emergence of new ponds in the surrounding areas. Such alterations in land use signify that growing human interference and consequent increases in exposed areas around the lake led to the possibility of higher rates of soil erosion than previously. The eroded materials were directly or indirectly added to the lake and resulted in its gradual siltation.

12.4.1.6 Eutrophication of the Rudrasagar Lake

Eutrophication refers to the excessive growth of algae and plankton in a water body like a lake mainly due to an oversupply of nutrients. Agricultural runoff, developmental activities, pollution from point and non-point sources, etc. are the human-related activities that increase the flow of nutrients in a lake and promote cultural eutrophication. Algae bloom reduces the amount of dissolved oxygen available for other plants and aquatic fauna and also minimizes the penetration of sunlight deep into the water. One of the major problems of the Rudrasagar Lake is the high amount

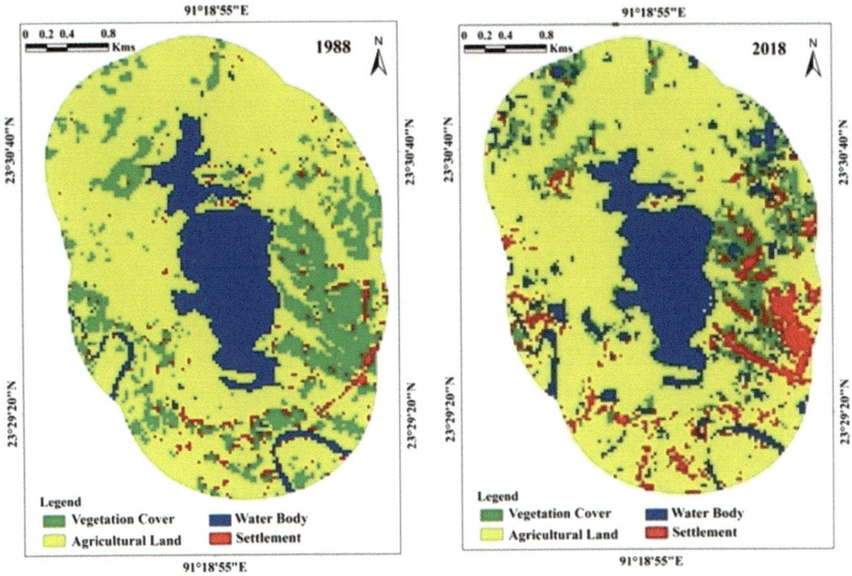

Fig. 12.9 Land use/land cover within 1 km buffer zone of the Rudrasagar Lake between 1988 and 2018

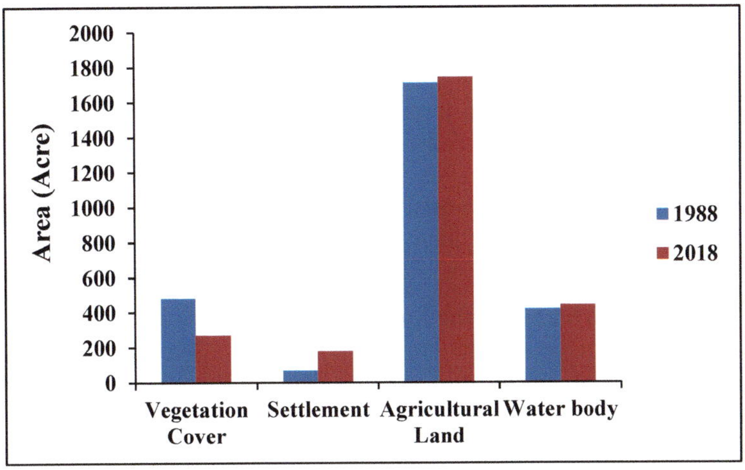

Fig. 12.10 Changing scenario of land use/land cover within a 1 km buffer zone around the Rudrasagar Lake between 1988 and 2018

Fig. 12.11 Eutrophication in the Rudrasagar Lake

of eutrophication (Fig. 12.11). Uncontrolled growth of alien invasive species such as water hyacinth and algae were identified as having caused loss of aquatic biodiversity (Deka 2010) in the lake.

12.4.2 Issues Arising Due to Growing Geo-Environmental Problems

The growing human population in the surroundings of the Rudrasagar Lake has resulted in the depletion of natural resources persisting there. Consequently, the amount of sediment supply to this lake has increased resulting in the gradual shrinkage of its areal extent. Moreover, several other natural and ecological problems have also arisen threatening the future existence of this pristine and important lake.

12.4.2.1 Areal Shrinkage of the Rudrasagar Lake

As discussed earlier, the amount of soil loss from the Rudrasagar Lake catchment has increased over time due to various human interventions. These eroded materials contribute to increasing the number of sediments that enter the lake. In the present study, the areal extent during January was considered to derive an insight about temporal variation in the spatial extension of the Rudrasagar Lake (Fig. 12.12). It was found that in January 1988, the area of the lake was about 373.26 acre, which decreased to 362.87 by January 2004 and to 354.03 acre by January 2018 (Fig. 12.13). Such a reduction is remarkable in the northern parts of the lake and constitutes a serious threat to the future existence of the lake.

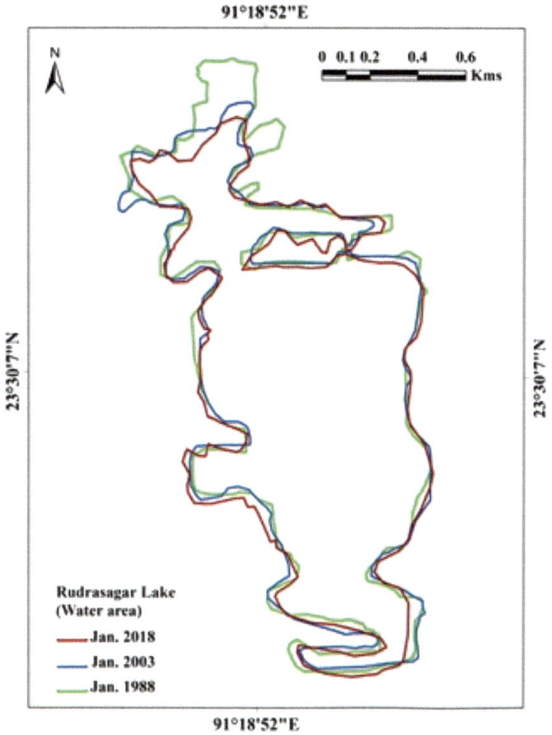

Fig. 12.12 Areal extent of the Rudrasagar Lake during of the years 1988, 2003, and 2018

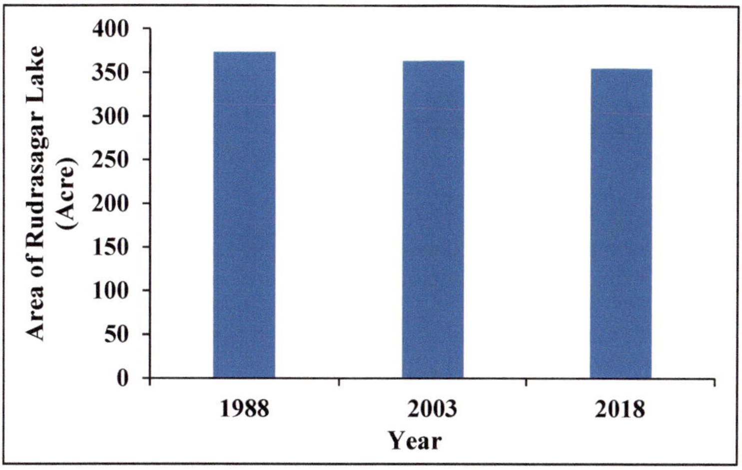

Fig. 12.13 Areal variation of the Rudrasagar Lake from the year 1988 to 2018

12.4.2.2 Inundation of the Surroundings

Due to the reduction of both depth as well as the areal extent, its capacity to hold the extra influx of water during the rainy period also diminished. Consequently, the surrounding areas were variously inundated (Fig. 12.14a).

A zone of inundation was identified from the satellite image of October 2017 (Fig. 14a). This zone falls under the actual area of the Rudrasagar Lake, the maximum portion (2015.99 acre) of which is used for agricultural purposes, especially during the dry seasons (Fig. 14b). A significant portion of this zone is also occupied by ponds (150.73 acre) and settlements (36.85 acre) (Table 12.3). These areas are inundated by lake water during torrential rainfall events causing economic loss to local residents.

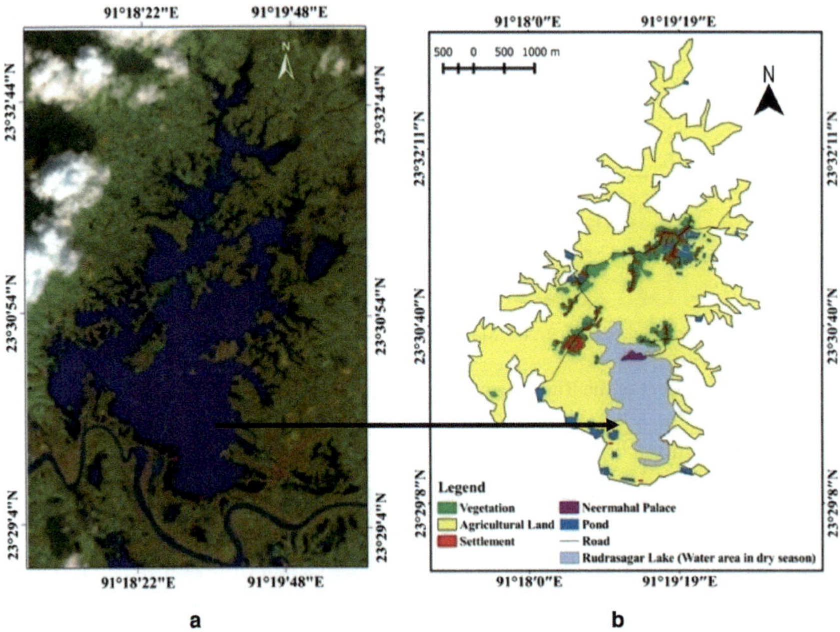

Fig. 12.14 **a** Inundated zone in the surroundings of the Rudrasagar Lake. **b** Land use of the inundated zone (*Source* Google Earth Imagery, 2017)

Table 12.3 Area under different land use/land cover classes within the inundated zone

Land use/land cover classes	Area /length (acres/km)
Agricultural land	2015.99
Pond	150.73
Vegetation	219.7
Settlement	36.85
Road Length	7.81 km

12.4.2.3 Declining Fish Production

Since the depth of the lake has decreased remarkably over the last few decades, fish production has naturally declined as well. During the survey, 89% of the respondents felt that the production of fish had gradually decreased (Fig. 12.15), causing the local fishermen to shift from fishing to other occupations as a livelihood strategy. Nearly 29% of the surveyed fishermen had already shifted to other occupation and some 38% were about to do so as well.

12.4.2.4 Effect on Migratory Birds

In the recent past, the Rudrasagar Lake was well known for migratory birds visiting it seasonally. However, during recent decades, on account of a decline in its areal extent as well as in terms of its physical quality, its appearance began to reduce perceptibly. During the field survey, most respondents (68%) agreed that birds like Babahans—*Swan (Cygnus),* Homokkocha—*Asian Openbil (Stork) (Anastomus oscitans),* Arkilla—*Koel (Eudynamys scolopaceus),* Hanikhowari (*Cormorant—Phalacrocorax sp.*) among other bird species had either stopped coming or their numbers had sharply declined. As such the ecological balance of the lake had come under stress.

12.5 Conclusion

Rudrasagar Lake, a Ramsar wetland of international significance is facing several geo-environmental problems which have led to its gradual degradation. Growing human interferences within the catchment area acts as a major driver in the gradual siltation and shrinkage of the lake. The foregoing analysis reveals that land use/land cover within the periphery of this lake was significantly modified during the past few

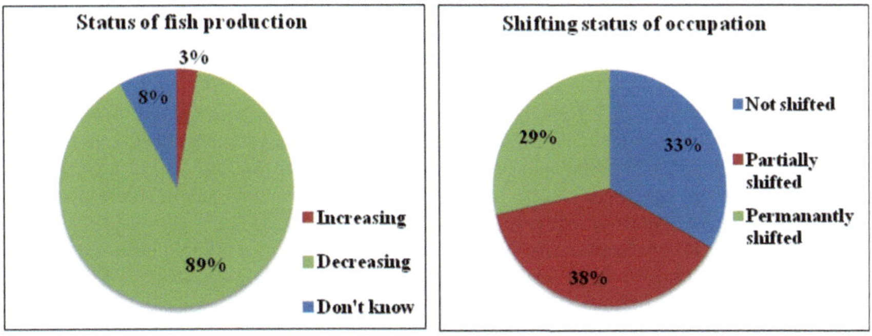

Fig. 12.15 Status of fish production and occupational shift of the fishermen

decades through the clearance of vegetal cover. From the application of the USLE model, it was observed that moderate, high, very high, and severe soil loss areas within the catchment of the Rudrasagar Lake had increased substantially during the 27-year period from 1990 to 2016. A significant reduction in the sediment load of the outflow channel from the Rudrasagar Lake compared to the inflow was observed. This indicates that a considerable portion of sediment that enters the Rudrasagar Lake settles down to the lakebed causing gradual siltation and degradation of the lake and consequent decreases in fish production. The latter has pressurized local fishermen to shift their occupation partially or completely. Thus, the formation and implementation of a suitable management strategy is the need of the hour. A social forestry program could be initiated in the very high to severe soil loss areas identified through the USLE model to reduce soil erosion from the catchment area of the Rudrasagar Lake. Moreover, any kind of human interference in the immediate periphery of this lake should be strictly prohibited. Dredging of the lakebed could be undertaken to improve its depth as a short-term measure. However, currently, in the absence of any management efforts, the very existence of this Ramsar wetland is, unfortunately, at stake.

References

Abir S (2014) Seasonal variations in Physico-chemical Cherracteristics of Rudrasagar Wetland—a Ramsar Site, Tripura, North East. India. Res J Chem Sci 4(1):31–40

Barman D, Mandal SC, Bhattacharjee P, Datta SP (2013) Status of Rudrasagar Lake (Ramsar Site) in Tripura. India. Environ Ecol 31(3):1320–1325

Choudhury AU (2008) Rudrasagar "A Potential IBA in Tripura in North-East India." Mistnet 9:4–5

Dabral PP, Baithuri N, Pandey A (2008) Soil erosion assessment in a hilly catchment of North Eastern India using USLE, GIS and remote sensing. Water Resour Manag 22:1783–1798

Deka S (2010) Conservation, restoration and management of Rudrasagar Lake (Tripura), Seminar proceedings, department of life science, Dibrugarh University

Durmaz F, Karakaya N, Evrendilek F (2016) Spatiotemporal change detection analysis of Turkish lake water surface area in response to anthropogenic ecosystem disturbances using long-term Landsat TM/ETM+ Data. J Ecosys Ecograph 6(2):188. https://doi.org/10.4172/2157.7625.1000188

Madhumathi T, Sharon Bercy J, Thangaperumal S (2018) Water quality assessment of Velacherry Lake using remote sensing and GIS techniques. Int J Sci Res 7(3):159–165. https://doi.org/10.21275/3031801

Maurya AK, Radhakrishnan KV, Kumar R (2020) Population characteristics and level of exploitation of Anabas testudineus (Bloch, 1792) in Rudrasagar Lake, a Ramsar site in North-eastern India. Indian J Mar Sci 49(2):298–302

Pal J, Pal M, Roy PK, Mazumdar A (2016) Quantitative assessment of water of Rudrasagar Lake, Tripura. India. Eur J Adv Eng Tech 3(3):45–48

Pal M, Samal NR, Roy MB, Roy PK (2015) Water quality index as a reliable indicator of water pollution level-A case study of Rudrasagar Lake, Tripura. Int J Innov Res Sci Eng Technol 4(8):7801–7806. https://doi.org/10.15680/IJIRSET.2015.0408138

Prasannakumar V, Vijith H, Abinod S, Geeth N (2012) Estimation of soil erosion risk within a small mountainous sub–watershed in Kerala, India, using revised universal soil loss equation (RUSLE) and geo–information technology. Geosci Front 3(2):209–215

Saraf M, Nair SJ (2015) Restoration and conservation of urban lakes. Int J Adv Eng Res Dev 2(6):315–323

GPSR Compliance

The European Union's (EU) General Product Safety Regulation (GPSR) is a set of rules that requires consumer products to be safe and our obligations to ensure this.

If you have any concerns about our products, you can contact us on ProductSafety@springernature.com

In case Publisher is established outside the EU, the EU authorized representative is:

Springer Nature Customer Service Center GmbH
Europaplatz 3
69115 Heidelberg, Germany

Batch number: 10732090

Printed by Printforce, the Netherlands